Peterson's
SAT
Math
Workbook

Peterson's
SAT
Math
Workbook

THOMSON

PETERSON'S

Australia • Canada • Mexico • Singapore • Spain • United Kingdom • United States

About Thomson Peterson's

Thomson Peterson's (www.petersons.com) is a leading provider of education information and advice, with books and online resources focusing on education search, test preparation, and financial aid. Its Web site offers searchable databases and interactive tools for contacting educational institutions, online practice tests and instruction, and planning tools for securing financial aid. Peterson's serves 110 million education consumers annually.

For more information, contact Peterson's, 2000 Lenox Drive, Lawrenceville, NJ 08648; 800-338-3282; or find us on the World Wide Web at www.petersons.com/about.

Editor: Wallie Walker Hammond; Production Editor: Alysha Bullock;
Manufacturing Manager: Judy Coleman; Composition Manager: Melissa Ignatowski;
Cover Design: Greg Wuttke

ISBN 13: 978-0-7689-1717-8
ISBN 10: 0-7689-1717-4

Printed in the United States of America

10 9 8 7 6 5 4   07 06

Contents

About the SAT

PURPOSE OF THE SAT

The SAT is a standardized exam used by many colleges and universities in the United States and Canada to help them make their admissions decisions. The test is developed and administered by Educational Testing Service (ETS) for the College Entrance Examination Board.

The SAT consists of two different types of exams designated SAT and SAT II. The SAT tests verbal and mathematical reasoning skills — your ability to understand what you read, to use language effectively, to reason clearly, and to apply fundamental mathematical principles to unfamiliar problems. SAT II tests mastery of specific subjects such as Chemistry or French or World History.

TAKING THE SAT

The SAT is offered on one Saturday morning in October, November, December, January, March, May, and June. When you apply to a college, find out whether it requires you to take the SAT and if so when scores are due. To make sure your scores arrive in time, sign up for a test date that's at least six weeks before the school's deadline for test scores.

Registration forms for the SAT are available in most high school guidance offices. You can also get registration forms and any other SAT information from:

College Board SAT Program
P.O. Box 6200
Princeton, NJ 08541-6200
609-771-7600

Monday through Friday, 8:30 a.m. to 9:30 p.m. Eastern Time

www.collegeboard.com

Along with your registration form you will receive a current *SAT Student Bulletin*. The bulletin includes all necessary information on procedures, exceptions and special arrangements, times and places, and fees.

FORMAT OF THE SAT

The SAT is a three-hour, 45-minute mostly multiple-choice examination divided into sections as shown in the chart below. One of the sections is experimental. Your score on the six nonexperimental sections is the score colleges use to evaluate your application.

The critical reading sections of the SAT use Sentence Completions to measure your knowledge of the meanings of words and your understanding of how parts of sentences go together, and Critical Reading questions (short and long passages) to measure your ability to read and think carefully about the information presented in passages.

The mathematical sections use Standard Multiple-Choice Math, Quantitative Comparisons, and Student-Produced Response Questions to test your knowledge of arithmetic, algebra, and geometry. Many of the formulas that you need will be given in the test instructions. You are not required to memorize them. SAT math questions are designed to test your skill in applying basic math principles you already know to unfamiliar situations.

The experimental section of SAT may test critical reading or mathematical reasoning, and **it can occur at any point during the test.** This section is used solely by the testmakers to try out questions for use in future tests. You won't know which section it is. So you'll have to do your best on all of the sections.

FORMAT OF A TYPICAL SAT

Section #/Content	Number of Questions	Time (approximate)
(1) Writing	1 essay	25 min.
(2) Mathematics Standard Multiple Choice	20	25 min.
(3)* "Wild Card" an Experimental Section **(Varies with test)**	varies	25 min.
(4) Critical Reading Sentence Completions	16 8	25 min.
(5) Writing Standard Multiple Choice	35	25 min.
(6) Mathematics Standard Multiple-choice Grid-Ins	8 10	25 min.
(7) Critical Reading Sentence Completions	19 5	25 min.
(8) Mathematics Standard Multiple Choice	16	20 min.
(9) Critical Reading Sentence Completions	13 6	20 min.
(10) Writing Standard Multiple Choice	10	10 min.

* Can occur in any section

THE SAT MATH QUESTIONS

The mathematical reasoning sections of the SAT test problem solving in numbers and operations, algebra I and II, geometry, statistics, probability, and data analysis using two question types:

- **Standard multiple-choice questions** give you a problem in arithmetic, algebra, or geometry. Then you choose the correct answer from the five choices.

- **Grid-Ins** do not give you answer choices. You have to compute the answer and then use the ovals on the answer sheet to fill in your solution.

Although calculators are not required to answer any SAT math questions, students are encouraged to bring a calculator to the test and to use it wherever it is helpful.

Mathematics tests your knowledge of arithmetic, algebra, and geometry. You are to select the correct solution to the problem from the five choices given.

Example:

If $(x + y)^2 = 17$, and $xy = 3$, then $x^2 + y^2 =$
- (A) 11
- (B) 14
- (C) 17
- (D) 20
- (E) 23

Solution:

The correct answer is (A).

$$(x + y)^2 = 17$$
$$(x + y)(x + y) = 17$$
$$x^2 + 2xy + y^2 = 17$$

Since $xy = 3$,

$$x^2 + 2(3) + y^2 = 17$$
$$x^2 + 6 + y^2 = 17$$
$$x^2 + y^2 = 11$$

Student-Produced Responses test your ability to solve mathematical problems when no choices are offered.

Example:

On a map having a scale of $\frac{1}{4}$ inch = 20 miles, how many inches should there be between towns that are 70 miles apart?

Solution:

The correct answer is $\frac{7}{8}$ or .875, depending upon whether you choose to solve the problem using fractions or decimals.

Using fractions	Using decimals
$\dfrac{\frac{1}{4}}{20} = \dfrac{x}{70}$	$\dfrac{.25}{20} = \dfrac{x}{70}$
$20x = \dfrac{70}{4}$	$20x = 17.5$
$x = \left(\dfrac{70}{4}\right)\left(\dfrac{1}{20}\right) = \dfrac{7}{8}$	$x = .875$

HOW TO USE THE ANSWER GRID

The answer grid for student-produced response (grid-ins) questions is similar to the grid used for your zip code on the personal information section of your answer sheet. An example of the answer grid is shown below.

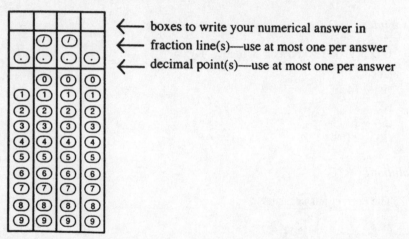

← boxes to write your numerical answer in
← fraction line(s)—use at most one per answer
← decimal point(s)—use at most one per answer

The open spaces above the grid are for you to write in the numerical value of your answer. The first row of ovals has only two ovals in the middle with a "/". These allow you to enter numbers in fractional form. Since a fraction must have both a numerator and a denominator, it is not possible that the leftmost or rightmost positions could have a "/". To protect you from yourself, there are no "/s" in those positions. The next row has decimal points. The horizontal bar separates the fraction lines and decimal points from the digits 0 to 9. Record your answers to grid-in questions according to the rules that follow.

GRID RULES

1. **Write your answer in the boxes at the top of the grid.**

   Technically this isn't required by the SAT. Realistically, it gives you something to follow as you fill in the ovals. Do it—it will help you.

2. **Mark the bubbles that correspond to the answer you entered in the boxes.**

   Mark one bubble per column. The machine that scores the test can only read the bubbles, so if you don't fill them in, you won't get credit. Just entering your answer in the boxes is not enough.

3. Start your answer in any column, if space permits.

Unused columns should be left blank. Don't put in zeroes. Look at this example:

Here are two ways to enter an answer of "150."

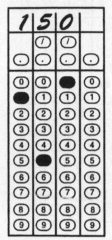 

4. Work with decimals or fractions.

An answer can be expressed as $\frac{3}{4}$ or as .75. Do not put a zero in front of a decimal that is less than 1. Just remember that you have only four spaces to work with and that a decimal point or a fraction slash uses up one of the spaces.

For decimal answers, be as accurate as possible but keep within the four spaces. Say you get an answer of .1777. Here are your options:

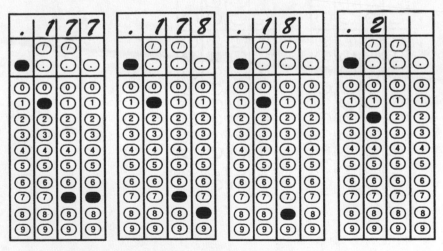

Fractions do not have to be simplified to simplest form unless they don't fit in the answer grid. For example, you can grid $\frac{4}{10}$, but you can't grid $\frac{12}{16}$ because you'd need five spaces. So, you would simplify it and grid $\frac{3}{4}$.

5. **Express a mixed number as a decimal or as an improper fraction.**

If you tried to grid $1\frac{3}{4}$, it would be read as $\frac{13}{4}$, which would give you a wrong answer. Instead you should grid this answer as 1.75 or as $\frac{7}{4}$.

6. **If more than one answer is possible, grid any one.**

Sometimes the problems in this section will have more than one correct answer. In such cases, choose one answer and grid it. For example, if a question asks for a prime number between 5 and 13, the answer could be 7 or 11. Grid 7 or grid 11, but don't put in both answers.

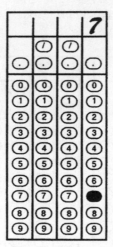 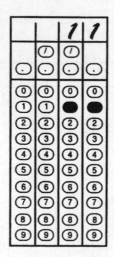

CALCULATORS AND THE SAT

Calculators are allowed on the SAT. You may bring to your exam any of the following types of calculators:

- four-function
- scientific
- graphing

You may not bring calculators of the following types:

- calculators with paper tape or printers
- laptop computers
- telephones with calculators
- "hand-held" microcomputers

Make sure that the calculator you bring is one you are thoroughly familiar with.

WHEN TO USE A CALCULATOR

No question requires the use of a calculator. For some questions a calculator may be helpful; for others it may be inappropriate. In general, the calculator may be useful for any question that involves arithmetic computations. Remember, though, that the calculator is only a tool. It can help you avoid inaccuracies in computation, but it cannot take the place of understanding how to set up and solve a mathematical problem.

Here is a sample problem for which a calculator would be useful:

Example:

The cost of two dozen apples is $3.60. At this rate, what is the cost of 10 apples?

(A) $1.75
(B) $1.60
(C) $1.55
(D) $1.50
(E) $1.25

Solution:

The correct answer is (D).
Make a ratio of apples to dollars:

$$\frac{\textbf{apples}}{\textbf{dollars}} : \frac{24}{3.60} = \frac{10}{x}$$

$$24x = 36$$

$$x = \frac{36}{24} = \$1.50$$

A calculator would be useful in solving this problem. Although the calculations are fairly simple, the calculator can improve your speed and accuracy.

Here is a problem for which a calculator would not be useful:

Example:

Joshua travels a distance of d miles in t - 6 hours. At this rate, how many miles will he travel in t^2 - 36 hours?

(A) $d(t + 6)$

(B) $d(t$ - $6)$

(C) $\dfrac{d}{t+6}$

(D) $\dfrac{d}{t-6}$

(E) $\dfrac{t+6}{d}$

Solution:

The correct answer is (A).

$$\text{rate} = \frac{\text{distance}}{\text{time}}$$

$$\text{Joshua's rate} = \frac{d}{t-6}$$

To calculate his new distance, use distance = rate × time

$$\text{Distance} = \left(\frac{d}{t-6}\right)(t^2 - 36)$$

$$= \left(\frac{d}{t-6}\right)(t+6)(t-6)$$

$$= d(t+6)$$

This is an algebra problem. Using a calculator would not be helpful.

SCORING THE SAT

Every correct answer is worth one point. If you leave an answer blank, you score no point. For incorrect answers to all verbal questions and to regular mathematics questions, you lose one-fourth of a point. For incorrect answers to quantitative comparisons, you lose one-third of a point. For incorrect answers to student-produced responses, there is no penalty. The penalties for wrong answers are intended to discourage random guessing.

Regardless of the number of questions on the test, all SAT scores are reported on a scale of 200 to 800. The scores are based on the nonexperimental sections and are broken down into separate math and verbal scores.

Five or six weeks after the exam, your scores will be sent to the colleges you have named on your registration form, to your high school, and to you.

HOW TO USE THIS BOOK

The math review that follows is designed as a self-teaching text to help you prepare for the mathematics sections of the SAT. At the beginning of each chapter, you will find a ten-question diagnostic test. Try this test before you read the chapter. Check your answers with the solutions provided at the end of the chapter. If you get eight to ten questions right, you may skip that chapter and go right on to the next diagnostic test at the beginning of the following chapter. Or you may prefer to skim the instructional material anyway, just for review, but not bother with the practice exercises. If you get five to seven questions right, you might do the practice exercises only in the sections dealing with problems you missed. If you get fewer than five questions right, you should work carefully through the entire chapter.

At the end of each chapter you will find a retest that is similar to the diagnostic test. After working through the chapter, you should do better on the retest. If not, go back and review any instructional material dealing with errors you made before proceeding to the next chapter.

Working diligently through each chapter in this manner will strengthen your weaknesses and prepare you to get your best score on the three Practice SAT Math Tests at the end of this book—and on your actual SAT.

Good luck.

Operations with Whole Numbers and Decimals

1

DIAGNOSTIC TEST

Directions: Work out each problem. Circle the letter that appears before your answer.

Answers are at the end of the chapter.

1. Find the sum of 683, 72, and 5429.

   (A) 5184
   (B) 6184
   (C) 6183
   (D) 6193
   (E) 6284

2. Subtract 417 from 804.

   (A) 287
   (B) 388
   (C) 397
   (D) 387
   (E) 288

3. Find the product of 307 and 46.

   (A) 3070
   (B) 14,082
   (C) 13,922
   (D) 13,882
   (E) 14,122

4. Divide 38,304 by 48.

   (A) 787
   (B) 798
   (C) 824
   (D) 1098
   (E) 1253

5. Add 6.43 + 46.3 + .346.

   (A) 14.52
   (B) 53.779
   (C) 53.392
   (D) 53.076
   (E) 1452

6. Subtract 81.763 from 145.1.

   (A) 64.347
   (B) 64.463
   (C) 63.463
   (D) 63.337
   (E) 63.347

7. Multiply 3.47 by 2.3.

   (A) 79.81
   (B) 7.981
   (C) 6.981
   (D) 7.273
   (E) 7.984

8. Divide 2.163 by .03.

   (A) 7210
   (B) 721
   (C) 72.1
   (D) 7.21
   (E) 0.721

9. Find $3 - 16 \div 8 + 4 \times 2$.

   (A) 9
   (B) $2\frac{1}{3}$
   (C) 10
   (D) 18
   (E) $\frac{2}{3}$

10. Which of the following is closest to $\frac{8317 \times 91}{217 \times .8}$?

    (A) 4
    (B) 40
    (C) 400
    (D) 4000
    (E) 40,000

In preparing for the mathematics section of your college entrance examination, it is most important to overcome any fear of mathematics. The level of this examination extends no further than relatively simple geometry. Most problems can be solved using only arithmetic. By reading this chapter carefully, following the sample problems, and then working on the practice problems in each section, you can review important concepts and vocabulary, as well as familiarize yourself with various types of questions. Since arithmetic is basic to any further work in mathematics, this chapter is extremely important and should not be treated lightly. By doing these problems carefully and reading the worked-out solutions, you can build the confidence needed to do well.

1. ADDITION OF WHOLE NUMBERS

In the process of addition, the numbers to be added are called *addends*. The answer is called the *sum*. In writing an addition problem, put one number underneath the other, being careful to keep columns straight with the units' digits one below the other. If you find a sum by adding from top to bottom, you can check it by adding from bottom to top.

Example:

Find the sum of 403, 37, 8314, and 5.

Solution:

$$
\begin{array}{r}
403 \\
37 \\
8314 \\
+\ \ \ 5 \\
\hline
8759
\end{array}
$$

Exercise 1

1. Find the sum of 360, 4352, 87, and 205.

   (A) 5013
   (B) 5004
   (C) 5003
   (D) 6004
   (E) 6013

2. Find the sum of 4321, 2143, 1234, and 3412.

   (A) 12,110
   (B) 11,011
   (C) 11,101
   (D) 11,111
   (E) 11,110

3. Add 56 + 321 + 8 + 42.

   (A) 427
   (B) 437
   (C) 517
   (D) 417
   (E) 527

4. Add 99 + 88 + 77 + 66 + 55.

   (A) 384
   (B) 485
   (C) 385
   (D) 375
   (E) 376

5. Add 1212 + 2323 + 3434 + 4545 + 5656.

   (A) 17,171
   (B) 17,170
   (C) 17,160
   (D) 17,280
   (E) 17,270

2. SUBTRACTION OF WHOLE NUMBERS

The number from which we subtract is called the *minuend*. The number which we take away is called the *subtrahend*. The answer in subtraction is called the *difference*.

If 5 is subtracted from 11, the minuend is 11, the subtrahend is 5, and the difference is 6.

Since we cannot subtract a larger number from a smaller one, we often must borrow in performing a subtraction. Remember that when we borrow, because of our base 10 number system, we reduce the digit to the left by 1, but increase the right-hand digit by 10.

Example:

$$\begin{array}{r} 54 \\ -\ 38 \\ \hline \end{array}$$

Since we cannot subtract 8 from 4, we borrow 1 from 5 and change the 4 to 14. We are really borrowing 1 from the tens column and, therefore, add 10 to the ones column. Then we can subtract.

Solution:

$$\begin{array}{r} 4\ ^14 \\ -\ 3\ 8 \\ \hline 1\ 6 \end{array}$$

Sometimes we must borrow across several columns.

Example:

$$\begin{array}{r} 503 \\ -\ 267 \\ \hline \end{array}$$

We cannot subtract 7 from 3 and cannot borrow from 0. Therefore we reduce the 5 by one and make the 0 into a 10. Then we can borrow 1 from the 10, making it a 9. This makes the 3 into 13.

Solution:

$$\begin{array}{r} 4\ ^10\ 3 \\ -\ 2\ 6\ 7 \\ \hline \end{array} \qquad \begin{array}{r} 4\ 9\ ^13 \\ -\ 2\ 6\ 7 \\ \hline 2\ 3\ 6 \end{array}$$

Exercise 2

1. Subtract 803 from 952.
   - (A) 248
   - (B) 148
   - (C) 249
   - (D) 149
   - (E) 147

2. From the sum of 837 and 415, subtract 1035.
   - (A) 217
   - (B) 216
   - (C) 326
   - (D) 227
   - (E) 226

3. From 1872 subtract the sum of 76 and 43.
   - (A) 1754
   - (B) 1838
   - (C) 1753
   - (D) 1839
   - (E) 1905

4. Find the difference between 732 and 237.
   - (A) 496
   - (B) 495
   - (C) 486
   - (D) 405
   - (E) 497

5. By how much does the sum of 612 and 315 exceed the sum of 451 and 283?
   - (A) 294
   - (B) 1661
   - (C) 293
   - (D) 197
   - (E) 193

3. MULTIPLICATION OF WHOLE NUMBERS

The answer to a multiplication problem is called the *product*. The numbers being multiplied are called factors of the product.

When multiplying by a number containing two or more digits, place value is extremely important when writing partial products. When we multiply 537 by 72, for example, we multiply first by 2 and then by 7. However, when we multiply by 7, we are really multiplying by 70 and therefore leave a 0 at the extreme right before we proceed with the multiplication.

Example:

$$
\begin{array}{r}
537 \\
\times\ 72 \\
\hline
1074 \\
+\ 37590 \\
\hline
38664
\end{array}
$$

If we multiply by a three-digit number, we leave one zero on the right when multiplying by the tens digit and two zeros on the right when multiplying by the hundreds digit.

Example:

$$
\begin{array}{r}
372 \\
\times\ 461 \\
\hline
372 \\
22320 \\
+\ 148800 \\
\hline
171492
\end{array}
$$

Exercise 3

Find the following products.

1. 526 multiplied by 317

    (A)   156,742
    (B)   165,742
    (C)   166,742
    (D)   166,748
    (E)   166,708

2. 8347 multiplied by 62

    (A)   517,514
    (B)   517,414
    (C)   517,504
    (D)   517,114
    (E)   617,114

3. 705 multiplied by 89

    (A)   11,985
    (B)   52,745
    (C)   62,705
    (D)   62,745
    (E)   15,121

4. 437 multiplied by 607

    (A)   265,259
    (B)   265,219
    (C)   265,359
    (D)   265,059
    (E)   262,059

5. 798 multiplied by 450

    (A)   358,600
    (B)   359,100
    (C)   71,820
    (D)   358,100
    (E)   360,820

4. DIVISION OF WHOLE NUMBERS

The number being divided is called the *dividend*. The number we are dividing by is called the *divisor*. The answer to the division is called the *quotient*. When we divide 18 by 6, 18 is the dividend, 6 is the divisor, and 3 is the quotient. If the quotient is not an integer, we have a *remainder*. The remainder when 20 is divided by 6 is 2, because 6 will divide 18 evenly, leaving a remainder of 2. The quotient in this case is $6\frac{2}{6}$. Remember that in writing the fractional part of a quotient involving a remainder, the remainder becomes the numerator and the divisor the denominator.

When dividing by a single-digit divisor, no long division procedures are needed. Simply carry the remainder of each step over to the next digit and continue.

Example:

$$6\overline{)5\ 8^4\ 3^1\ 4^2\ 4} = 9\ 7\ 2\ 4$$

Exercise 4

1. Divide 391 by 23.

   (A) 170
   (B) 16
   (C) 17
   (D) 18
   (E) 180

2. Divide 49,523,436 by 9.

   (A) 5,502,605
   (B) 5,502,514
   (C) 5,502,604
   (D) 5,502,614
   (E) 5,502,603

3. Find the remainder when 4832 is divided by 15.

   (A) 1
   (B) 2
   (C) 3
   (D) 4
   (E) 5

4. Divide 42,098 by 7.

   (A) 6014
   (B) 6015
   (C) 6019
   (D) 6011
   (E) 6010

5. Which of the following is the quotient of 333,180 and 617?

   (A) 541
   (B) 542
   (C) 549
   (D) 540
   (E) 545

5. ADDITION OR SUBTRACTION OF DECIMALS

The most important thing to watch for in adding or subtracting decimals is to keep all decimal points underneath one another. The proper placement of the decimal point in the answer will be in line with all the decimal points above.

Example:

Find the sum of 8.4, .37, and 2.641

Solution:

$$
\begin{array}{r}
8.4 \\
.37 \\
+\ 2.641 \\
\hline
11.411
\end{array}
$$

Example:

From 48.3 subtract 27.56

Solution:

$$
\begin{array}{r}
{}^{7}\ {}^{12}\ {}^{1} \\
4\,8.\cancel{3}\,0 \\
-\ 2\,7.5\,6 \\
\hline
2\,0.7\,4
\end{array}
$$

In subtraction, the upper decimal must have as many decimal places as the lower, so we must fill in zeros where needed.

Exercise 5

1. From the sum of .65, 4.2, 17.63, and 8, subtract 12.7.

    (A) 9.78
    (B) 17.68
    (C) 17.78
    (D) 17.79
    (E) 18.78

2. Find the sum of .837, .12, 52.3, and .354.

    (A) 53.503
    (B) 53.611
    (C) 53.601
    (D) 54.601
    (E) 54.611

3. From 561.8 subtract 34.75.

    (A) 537.05
    (B) 537.15
    (C) 527.15
    (D) 527.04
    (E) 527.05

4. From 53.72 subtract the sum of 4.81 and 17.5.

    (A) 31.86
    (B) 31.41
    (C) 41.03
    (D) 66.41
    (E) 41.86

5. Find the difference between 100 and 52.18.

    (A) 37.82
    (B) 47.18
    (C) 47.92
    (D) 47.82
    (E) 37.92

6. MULTIPLICATION OF DECIMALS

In multiplying decimals, we proceed as we do with integers, using the decimal points only as an indication of where to place a decimal point in the product. The number of decimal places in the product is equal to the sum of the number of decimal places in the numbers being multiplied.

Example:

> Multiply .375 by .42

Solution:

$$
\begin{array}{r}
.375 \\
\times \ \ .42 \\
\hline
750 \\
+ \ \ 15000 \\
\hline
.15750
\end{array}
$$

Since the first number being multiplied contains three decimal places and the second number contains two decimal places, the product will contain five decimal places.

To multiply a decimal by 10, 100, 1000, etc., we need only to move the decimal point to the right the proper number of places. In multiplying by 10, move one place to the right (10 has one zero), by 100 move two places to the right (100 has two zeros), by 1000 move three places to the right (1000 has three zeros), and so forth.

Example:

> The product of .837 and 100 is 83.7

Exercise 6

Find the following products.

1. $437 \times .24 =$

   (A) 1.0488
   (B) 10.488
   (C) 104.88
   (D) 1048.8
   (E) 10,488

2. $5.06 \times .7 =$

   (A) .3542
   (B) .392
   (C) 3.92
   (D) 3.542
   (E) 35.42

3. $83 \times 1.5 =$

   (A) 12.45
   (B) 49.8
   (C) 498
   (D) 124.5
   (E) 1.245

4. $.7314 \times 100 =$

   (A) .007314
   (B) .07314
   (C) 7.314
   (D) 73.14
   (E) 731.4

5. $.0008 \times 4.3 =$

   (A) .000344
   (B) .00344
   (C) .0344
   (D) 0.344
   (E) 3.44

7. DIVISION OF DECIMALS

When dividing by a decimal, always change the decimal to a whole number by moving the decimal point to the end of the divisor. Count the number of places you have moved the decimal point and move the dividend's decimal point the same number of places. The decimal point in the quotient will be directly above the one in the dividend.

Example:

Divide 2.592 by .06

Solution:

$$.06\overline{)2.592} \quad \frac{43.2}{}$$

To divide a decimal by 10, 100, 1000, etc., we move the decimal point the proper number of places to the *left*. The number of places to be moved is always equal to the number of zeros in the divisor.

Example:

Divide 43.7 by 1000

Solution:

The decimal point must be moved three places (there are three zeros in 1000) to the left. Therefore, our quotient is .0437

Sometimes division can be done in fraction form. Always remember to move the decimal point to the end of the divisor (denominator) and then the same number of places in the dividend (numerator).

Example:

Divide: $\frac{.0175}{.05} = \frac{1.75}{5} = .35$

Exercise 7

1. Divide 4.3 by 100.
   (A) .0043
   (B) 0.043
   (C) 0.43
   (D) 43
   (E) 430

2. Find the quotient when 4.371 is divided by .3.
   (A) 0.1457
   (B) 1.457
   (C) 14.57
   (D) 145.7
   (E) 1457

3. Divide .64 by .4.
   (A) .0016
   (B) 0.016
   (C) 0.16
   (D) 1.6
   (E) 16

4. Find $.12 \div \frac{2}{.5}$.
   (A) 4.8
   (B) 48
   (C) .03
   (D) 0.3
   (E) 3

5. Find $\frac{10.2}{.03} \div \frac{1.7}{.1}$.
   (A) .02
   (B) 0.2
   (C) 2
   (D) 20
   (E) 200

8. THE LAWS OF ARITHMETIC

Addition and multiplication are *commutative* operations, as the order in which we add or multiply does not change an answer.

Example:

$$4 + 7 = 7 + 4$$
$$5 \cdot 3 = 3 \cdot 5$$

Subtraction and division are not commutative, as changing the order does change the answer.

Example:

$$5 - 3 \neq 3 - 5$$
$$20 \div 5 \neq 5 \div 20$$

Addition and multiplication are *associative*, as we may group in any manner and arrive at the same answer.

Example:

$$(3 + 4) + 5 = 3 + (4 + 5)$$
$$(3 \cdot 4) \cdot 5 = 3 \cdot (4 \cdot 5)$$

Subtraction and division are not associative, as regrouping changes an answer.

Example:

$$(5 - 4) - 3 \neq 5 - (4 - 3)$$
$$(100 \div 20) \div 5 \neq 100 \div (20 \div 5)$$

Multiplication is *distributive* over addition. If a sum is to be multiplied by a number, we may multiply each addend by the given number and add the results. This will give the same answer as if we had added first and then multiplied.

Example:

$$3(5 + 2 + 4) \text{ is either } 15 + 6 + 12 \text{ or } 3(11).$$

The *identity for addition* is 0 since any number plus 0, or 0 plus any number, is equal to the given number.

The *identity for multiplication* is 1 since any number times 1, or 1 times any number, is equal to the given number.

There are no identity elements for subtraction or division. Although $5 - 0 = 5$, $0 - 5 \neq 5$. Although $8 \div 1 = 8$, $1 \div 8 \neq 8$.

When several operations are involved in a single problem, parentheses are usually included to make the order of operations clear. If there are no parentheses, multiplication and division are always performed prior to addition and subtraction.

Example:

Find $5 \cdot 4 + 6 \div 2 - 16 \div 4$

Solution:

The + and – signs indicate where groupings should begin and end. If we were to insert parentheses to clarify operations, we would have $(5 \cdot 4) + (6 \div 2) - (16 \div 4)$, giving $20 + 3 - 4 = 19$.

Exercise 8

1. Find $8 + 4 \div 2 + 6 \cdot 3 - 1$.
   - (A) 35
   - (B) 47
   - (C) 43
   - (D) 27
   - (E) 88

2. $16 \div 4 + 2 \cdot 3 + 2 - 8 \div 2$.
   - (A) 6
   - (B) 8
   - (C) 2
   - (D) 4
   - (E) 10

3. Match each illustration in the left-hand column with the law it illustrates from the right-hand column.

   a. $475 \cdot 1 = 475$ u. Identity for Addition
   b. $75 + 12 = 12 + 75$ v. Associative Law of Addition
   c. $32(12 + 8) = 32(12) + 32(8)$ w. Associative Law of Multiplication
   d. $378 + 0 = 378$ x. Identity for Multiplication
   e. $(7 \cdot 5) \cdot 2 = 7 \cdot (5 \cdot 2)$ y. Distributive Law of Multiplication
                                                     over Addition
                                              z. Commutative Law of Addition

9. ESTIMATING ANSWERS

On a competitive examination, where time is an important factor, it is essential that you be able to estimate an answer. Simply round off all answers to the nearest multiples of 10 or 100 and estimate with the results. On multiple-choice tests, this should enable you to pick the correct answer without any time-consuming computation.

Example:

The product of 498 and 103 is approximately

(A) 5000
(B) 500,000
(C) 50,000
(D) 500
(E) 5,000,000

Solution:

498 is about 500. 103 is about 100. Therefore the product is about (500) (100) or 50,000 (just move the decimal point two places to the right when multiplying by 100). Therefore, the correct answer is (C).

Example:

Which of the following is closest to the value of $1831 \cdot \frac{710}{2314}$?

(A) 83
(B) 425
(C) 1600
(D) 3140
(E) 6372

Solution:

Estimating, we have $\left(\frac{(5000)(700)}{2000} \right)$. Dividing numerator and denominator by 1000, we have $\frac{5(700)}{2}$ or $\frac{3500}{2}$, which is about 1750. Therefore, we choose answer (C).

Exercise 9

Choose the answer closest to the exact value of each of the following problems. Use estimation in your solutions. No written computation should be needed. Circle the letter before your answer.

1. $\frac{483+1875}{119}$

(A) 2
(B) 10
(C) 20
(D) 50
(E) 100

2. $\frac{6017 \cdot 312}{364+618}$

(A) 18
(B) 180
(C) 1800
(D) 18,000
(E) 180,000

3. $\frac{783+491}{1532-879}$

(A) .02
(B) .2
(C) 2
(D) 20
(E) 200

RETEST

1. Find the sum of 86, 4861, and 205.

   (A) 5142
   (B) 5132
   (C) 5152
   (D) 5052
   (E) 4152

2. From 803 subtract 459.

   (A) 454
   (B) 444
   (C) 354
   (D) 344
   (E) 346

3. Find the product of 65 and 908.

   (A) 59,020
   (B) 9988
   (C) 58,920
   (D) 58,020
   (E) 59,920

4. Divide 66,456 by 72.

   (A) 903
   (B) 923
   (C) 911
   (D) 921
   (E) 925

5. Find the sum of .361 + 8.7 + 43.17.

   (A) 52.078
   (B) 51.538
   (C) 51.385
   (D) 52.161
   (E) 52.231

6. Subtract 23.17 from 50.9.

   (A) 26.92
   (B) 27.79
   (C) 27.73
   (D) 37.73
   (E) 37.79

7. Multiply 8.35 by .43.

   (A) 3.5805
   (B) 3.5905
   (C) 3.5915
   (D) 35.905
   (E) .35905

8. Divide 2.937 by .11.

   (A) .267
   (B) 2.67
   (C) 26.7
   (D) 267
   (E) 2670

9. Find $8 + 10 \div 2 + 4 \cdot 2 - 21 \div 7$.

   (A) 17
   (B) 23
   (C) 18
   (D) 14
   (E) $\dfrac{5}{7}$

10. Which of the following is closest to $\dfrac{\frac{2875+932}{5817}}{29}$?

   (A) .02
   (B) .2
   (C) 2
   (D) 20
   (E) 200

SOLUTIONS TO PRACTICE EXERCISES

Diagnostic Test

1. (B)
$$
\begin{array}{r}
683 \\
72 \\
+\ 5429 \\
\hline
6184
\end{array}
$$

2. (D)
$$
\begin{array}{r}
\overset{7\ \ 9}{8\,\cancel{0}\,^{1}4} \\
-4\ 1\ 7 \\
\hline
3\ 8\ 7
\end{array}
$$

3. (E)
$$
\begin{array}{r}
307 \\
\times\ 46 \\
\hline
1842 \\
12280 \\
\hline
14{,}122
\end{array}
$$

4. (B)
$$
\begin{array}{r}
798 \\
48\overline{)38304} \\
\underline{336} \\
470 \\
\underline{432} \\
384 \\
384
\end{array}
$$

5. (D)
$$
\begin{array}{r}
6.43 \\
46.3 \\
+\ .346 \\
\hline
53.076
\end{array}
$$

6. (D)
$$
\begin{array}{r}
14\overset{4}{\cancel{5}}.\overset{10}{\cancel{1}}\,\overset{9}{\cancel{0}}\,^{1}0 \\
-\ \ 81.763 \\
\hline
63.337
\end{array}
$$

7. (B)
$$
\begin{array}{r}
3.47 \\
\times 2.3 \\
\hline
1041 \\
6940 \\
\hline
7.981
\end{array}
$$

8. (C)
$$
\begin{array}{r}
72.1 \\
.03\overline{)2.163}
\end{array}
$$

9. (A) $3 - (16 \div 8) + (4 \times 2) = 3 - 2 + 8 = 9$

10. (D) Estimate $\dfrac{8000 \cdot 100}{200 \cdot 1} = 4000$

Exercise 1

1. (B)
$$
\begin{array}{r}
360 \\
4352 \\
87 \\
+205 \\
\hline
5004
\end{array}
$$

2. (E)
$$
\begin{array}{r}
4321 \\
2143 \\
1234 \\
+3412 \\
\hline
11{,}110
\end{array}
$$

3. (A)
$$
\begin{array}{r}
56 \\
321 \\
8 \\
+\ 42 \\
\hline
427
\end{array}
$$

4. (C)
$$
\begin{array}{r}
99 \\
88 \\
77 \\
66 \\
+\ 55 \\
\hline
385
\end{array}
$$

5. (B)
$$
\begin{array}{r}
1212 \\
2323 \\
3434 \\
4545 \\
+5656 \\
\hline
17{,}170
\end{array}
$$

Exercise 2

1. (D)
$$9\overset{4}{\cancel{5}}{}^{1}2$$
$$-8\ 0\ 3$$
$$\overline{1\ 4\ 9}$$

2. (A)
$$8\ 3\ 7$$
$$+4\ 1\ 5$$
$$\overline{1\ 2\overset{4}{\cancel{5}}{}^{1}2}$$
$$-1\ 0\ 3\ 5$$
$$\overline{2\ 1\ 7}$$

3. (C)
$$76$$
$$+\ \ 43$$
$$\overline{119}$$

$$18\overset{6}{\cancel{7}}{}^{1}2$$
$$-1\ 1\ 9$$
$$\overline{17\ 5\ 3}$$

4. (B)
$$7\overset{6\ 12}{\cancel{3}}{}^{1}2$$
$$-2\ 3\ 7$$
$$\overline{4\ 9\ 5}$$

5. (E)
$$612$$
$$+315$$
$$\overline{927}$$
$$451$$
$$+\ 283$$
$$\overline{734}$$

$$\overset{8}{\cancel{9}}{}^{1}27$$
$$-7\ 34$$
$$\overline{1\ 93}$$

Exercise 3

1. (C)
$$526$$
$$\times\ \ \ 317$$
$$\overline{3682}$$
$$5260$$
$$\underline{157800}$$
$$166,742$$

2. (A)
$$8347$$
$$\times\ \ \ \ 62$$
$$\overline{16694}$$
$$\underline{500820}$$
$$517,514$$

3. (D)
$$705$$
$$\times\ \ \ 89$$
$$\overline{6345}$$
$$\underline{56400}$$
$$62,745$$

4. (A)
$$437$$
$$\times\ \ \ 607$$
$$\overline{3059}$$
$$\underline{262200}$$
$$265,259$$

5. (B)
$$798$$
$$\times\ \ \ 450$$
$$\overline{39900}$$
$$\underline{319200}$$
$$359,100$$

Exercise 4

1. (C)
$$23\overline{)391}^{\,17}$$
$$\underline{23}$$
$$161$$
$$161$$

2. (C)
$$9\overline{)49,523,436}^{\,5,502,604}$$

3. (B)
$$15\overline{)4832}^{\,322} \quad \text{Remainder } 2$$
$$\underline{45}$$
$$33$$
$$\underline{30}$$
$$32$$
$$\underline{30}$$
$$2$$

4. (A)
$$7\overline{)42098}^{\,6014}$$

5. (D) Since the quotient, when multiplied by 617, must give 333,180 as an answer, the quotient must end in a number which, when multiplied by 617, will end in 0. This can only be (D), since 617 times (A) would end in 7, (B) would end in 4, (C) in 3, and (E) in 5.

Exercise 5

1. (C)
$$.65$$
$$4.2$$
$$17.63$$
$$+\ \underline{8.}$$
$$30.48$$

$$\overset{2}{\cancel{3}}\,\overset{9}{\cancel{0}}.{}^{1}48$$
$$-1\,2\,.70$$
$$1\,7\,.78$$

2. (B)
$$.837$$
$$.12$$
$$52.3$$
$$+\ \underline{.354}$$
$$53.611$$

3. (E)
$$5\overset{5}{\cancel{6}}{}^{1}1\overset{8}{\cancel{7}}{}^{1}0$$
$$-\ \underline{3\,4.7\,5}$$
$$5\,2\,7.0\,5$$

4. (B)
$$4.81 \qquad 53.72$$
$$\underline{+17.5} \qquad \underline{-22.31}$$
$$22.31 \qquad 31.41$$

5. (D)
$$\overset{9}{\cancel{10}}\,\overset{1\,9}{\cancel{0}}.\overset{1\,9}{\cancel{0}}{}^{1}0$$
$$-\underline{5\,2\,.1\,8}$$
$$4\,7\,.8\,2$$

Exercise 6

1. (C)
$$
\begin{array}{r}
437 \\
\times\ .24 \\
\hline
1748 \\
8740 \\
\hline
104.88
\end{array}
$$

2. (D)
$$
\begin{array}{r}
5.06 \\
\times\ \ .7 \\
\hline
3.542
\end{array}
$$

3. (D)
$$
\begin{array}{r}
83 \\
\times\ 1.5 \\
\hline
415 \\
830 \\
\hline
124.5
\end{array}
$$

4. (D)
$$
\begin{array}{r}
.7314 \\
\times 100 \\
\hline
73.14
\end{array}
$$
Just move the decimal point two places to the right.

5. (B)
$$
\begin{array}{r}
.0008 \\
\times\ \ 4.3 \\
\hline
24 \\
320 \\
\hline
.00344
\end{array}
$$

Exercise 7

1. (B) Just move decimal point two places to left, giving .043 as the answer.

2. (C)
$$
.3\overline{)4.371} = 14.57
$$

3. (D)
$$
.4\overline{).64} = 1.6
$$

4. (C) $.12 \div \dfrac{2.0}{.5} = .12 \div 4 = .03$

5. (D) $\dfrac{10.20}{.03} \div \dfrac{1.7}{.1} = 340 \div 17 = 20$

Exercise 8

1. (D) $8 + (4 \div 2) + (6 \cdot 3) - 1 =$
   $8 + 2 + 18 - 1 = 27$

2. (B) $(16 \div 4) + (2 \cdot 3) + 2 - (8 \div 2) =$
   $4 + 6 + 2 - 4 = 8$

3. $(a, x)(b, z)(c, y)(d, u)(e, w)$

Exercise 9

1. (C) Estimate $\dfrac{500 + 2000}{100} = \dfrac{2500}{100} = 25$, closest to 20

2. (C) Estimate $\dfrac{6000 \cdot 300}{400 + 600} = \dfrac{1,800,000}{1000} = 1800$

3. (C) Estimate $\dfrac{800 + 500}{1500 - 900} = \dfrac{1300}{600} =$ about 2

Retest

1. (C)
$$
\begin{array}{r}
86 \\
4861 \\
+205 \\
\hline
5152
\end{array}
$$

2. (D)
$$
\begin{array}{r}
{}^{7}\;{}^{9} \\
8\,\cancel{0}\,{}^{1}3 \\
-4\,5\,9 \\
\hline
3\,4\,4
\end{array}
$$

3. (A)
$$
\begin{array}{r}
908 \\
\times\;\;65 \\
\hline
4540 \\
54480 \\
\hline
59{,}020
\end{array}
$$

4. (B)
$$
\begin{array}{r}
923 \\
72\overline{)66456} \\
\underline{648} \\
165 \\
\underline{144} \\
216 \\
216
\end{array}
$$

5. (E)
$$
\begin{array}{r}
.361 \\
8.7 \\
+43.17 \\
\hline
52.231
\end{array}
$$

6. (C)
$$
\begin{array}{r}
{}^{4}\;{}^{8} \\
\cancel{5}\,{}^{1}0.\cancel{9}\,{}^{1}0 \\
-2\;\;3.1\;7 \\
\hline
2\;7.7\;3
\end{array}
$$

7. (B)
$$
\begin{array}{r}
8.35 \\
\times\;.43 \\
\hline
2505 \\
33400 \\
\hline
3.5905
\end{array}
$$

8. (C)
$$
\begin{array}{r}
26.7 \\
.11\overline{)2.937} \\
\underline{22} \\
73 \\
\underline{66} \\
77 \\
77
\end{array}
$$

9. (C) $8 + (10 \div 2) + (4 \bullet 2) - (21 \div 7) =$
    $8 + 5 + 8 - 3 = 18$

10. (A) Estimate $\dfrac{\dfrac{3000+1000}{6000}}{30} = \dfrac{4000}{180{,}000} = .0\overline{2}$, which is

    closest to .02.

Operations with Fractions

2

DIAGNOSTIC TEST

Directions: Work out each problem. Circle the letter that appears before your answer.

Answers are at the end of the chapter.

1. The sum of $\frac{3}{5}$, $\frac{2}{3}$, and $\frac{1}{4}$ is

   (A) $\frac{1}{2}$

   (B) $\frac{27}{20}$

   (C) $\frac{3}{2}$

   (D) $\frac{91}{60}$

   (E) $1\frac{5}{12}$

4. $\frac{5}{6} \div \left(\frac{4}{3} \cdot \frac{5}{4} \right)$ is equal to

   (A) 2

   (B) $\frac{50}{36}$

   (C) $\frac{1}{2}$

   (D) $\frac{36}{50}$

   (E) $\frac{7}{12}$

2. Subtract $\frac{3}{4}$ from $\frac{9}{10}$.

   (A) $\frac{3}{20}$

   (B) 1

   (C) $\frac{3}{5}$

   (D) $\frac{3}{40}$

   (E) $\frac{7}{40}$

5. Subtract $32\frac{3}{5}$ from 57.

   (A) $24\frac{2}{5}$

   (B) $25\frac{3}{5}$

   (C) $25\frac{2}{5}$

   (D) $24\frac{3}{5}$

   (E) $24\frac{1}{5}$

3. The number 582,354 is divisible by

   (A) 4
   (B) 5
   (C) 8
   (D) 9
   (E) 10

6. Divide $4\frac{1}{2}$ by $1\frac{1}{8}$.

    (A) $\frac{1}{4}$

    (B) 4

    (C) $\frac{8}{9}$

    (D) $\frac{9}{8}$

    (E) $3\frac{1}{2}$

7. Which of the following fractions is the largest?

    (A) $\frac{1}{2}$

    (B) $\frac{11}{16}$

    (C) $\frac{5}{8}$

    (D) $\frac{21}{32}$

    (E) $\frac{3}{4}$

8. Which of the following fractions is closest to $\frac{2}{3}$?

    (A) $\frac{11}{15}$

    (B) $\frac{7}{10}$

    (C) $\frac{4}{5}$

    (D) $\frac{1}{2}$

    (E) $\frac{5}{6}$

9. Simplify $\dfrac{4-\dfrac{9}{10}}{\dfrac{2}{3}+\dfrac{1}{2}}$.

    (A) $\frac{93}{5}$

    (B) $\frac{93}{35}$

    (C) $\frac{147}{35}$

    (D) $\frac{147}{5}$

    (E) $\frac{97}{35}$

10. Find the value of $\dfrac{\dfrac{1}{a}+\dfrac{1}{b}}{\dfrac{1}{a}-\dfrac{1}{b}}$ when $a=3, b=4$.

    (A) 7

    (B) 2

    (C) 1

    (D) $\frac{1}{7}$

    (E) $\frac{2}{7}$

1. ADDITION AND SUBTRACTION

To add or subtract fractions, they must have the same *denominator*. To add several fractions, this common denominator will be the least number into which each given denominator will divide evenly.

Example:

Add $\dfrac{1}{2}+\dfrac{1}{3}+\dfrac{1}{4}+\dfrac{1}{5}$

Solution:

The common denominator must contain two factors of 2 to accommodate the 4, and also a factor of 3 and one of 5. That makes the least common denominator 60. Rename each fraction to have 60 as the denominator by dividing the given denominator into 60 and multiplying the quotient by the given numerator.

$$\frac{30}{60}+\frac{20}{60}+\frac{15}{60}+\frac{12}{60}=\frac{77}{60}=1\frac{17}{60}$$

When only two fractions are being added, a shortcut method can be used: $\dfrac{a}{b}+\dfrac{c}{d}=\dfrac{ad+bc}{bd}$. That is, in order to add two fractions, add the two cross products and place this sum over the product of the given denominators.

Example:

$\dfrac{4}{5}+\dfrac{7}{12}$

Solution:

$$\frac{4(12)+5(7)}{5(12)}=\frac{48+35}{60}=\frac{83}{60}=1\frac{23}{60}$$

A similar shortcut applies to the subtraction of two fractions:

$$\frac{a}{b}-\frac{c}{d}=\frac{ad-bc}{bd}$$

Example:

$$\frac{4}{5}-\frac{7}{12}=\frac{4(12)-5(7)}{5(12)}=\frac{48-35}{60}=\frac{13}{60}$$

Exercise 1

Work out each problem. Circle the letter that appears before your answer.

1. The sum of $\frac{1}{2}+\frac{2}{3}+\frac{3}{4}$ is

    (A) $\frac{6}{9}$

    (B) $\frac{23}{12}$

    (C) $\frac{23}{36}$

    (D) $\frac{6}{24}$

    (E) $2\frac{1}{3}$

2. The sum of $\frac{5}{17}$ and $\frac{3}{15}$ is

    (A) $\frac{126}{255}$

    (B) $\frac{40}{255}$

    (C) $\frac{8}{32}$

    (D) $\frac{40}{32}$

    (E) $\frac{126}{265}$

3. From the sum of $\frac{3}{4}$ and $\frac{5}{6}$ subtract the sum of $\frac{1}{4}$ and $\frac{2}{3}$.

    (A) 2

    (B) $\frac{1}{2}$

    (C) $\frac{36}{70}$

    (D) $\frac{2}{3}$

    (E) $\frac{5}{24}$

4. Subtract $\frac{3}{5}$ from $\frac{9}{11}$.

    (A) $-\frac{12}{55}$

    (B) $\frac{12}{55}$

    (C) 1

    (D) $\frac{3}{8}$

    (E) $\frac{3}{4}$

5. Subtract $\frac{5}{8}$ from the sum of $\frac{1}{4}$ and $\frac{2}{3}$.

    (A) 2

    (B) $\frac{3}{2}$

    (C) $\frac{11}{24}$

    (D) $\frac{8}{15}$

    (E) $\frac{7}{24}$

2. MULTIPLICATION AND DIVISION

In multiplying fractions, always try to divide out any common factor of any denominator with any numerator to keep your numbers as small as possible. Remember that if all numbers divide out in the numerator, you are left with a numerator of 1. The same goes for the denominator. If all numbers in both numerator and denominator divide out, you are left with $\frac{1}{1}$ or 1.

Example:

Multiply $\frac{3}{5} \cdot \frac{15}{33} \cdot \frac{11}{45}$

Solution:

$$\frac{\cancel{3}}{\cancel{5}} \cdot \frac{\cancel{15}}{\cancel{33}} \cdot \frac{\cancel{11}}{\cancel{45}} = \frac{1}{15}$$

In dividing fractions, we multiply by the multiplicative inverse.

Example:

Divide $\frac{5}{18}$ by $\frac{5}{9}$

Solution:

$$\frac{\cancel{5}}{\cancel{18}} \cdot \frac{\cancel{9}}{\cancel{5}} = \frac{1}{2}$$

Exercise 2

Work out each problem. Circle the letter that appears before your answer.

1. Find the product of $\frac{3}{2}$, 6, $\frac{4}{9}$, and $\frac{1}{12}$.

    (A) 3

    (B) $\frac{1}{3}$

    (C) $\frac{14}{23}$

    (D) $\frac{1}{36}$

    (E) $\frac{5}{12}$

2. Find $\frac{7}{8} \cdot \frac{2}{3} \div \frac{1}{8}$.

    (A) $\frac{3}{14}$

    (B) $\frac{7}{96}$

    (C) $\frac{21}{128}$

    (D) $\frac{14}{3}$

    (E) $\frac{8}{3}$

3. $\frac{3}{5} \div \left(\frac{1}{2} \cdot \frac{3}{10} \right)$ is equal to

    (A) 4

    (B) $\frac{1}{4}$

    (C) $\frac{12}{5}$

    (D) $\frac{5}{12}$

    (E) $\frac{12}{15}$

4. Find $\frac{2}{3}$ of $\frac{7}{12}$.

    (A) $\frac{7}{8}$

    (B) $\frac{7}{9}$

    (C) $\frac{8}{7}$

    (D) $\frac{8}{9}$

    (E) $\frac{7}{18}$

5. Divide 5 by $\frac{5}{12}$.

    (A) $\frac{25}{12}$

    (B) $\frac{1}{12}$

    (C) $\frac{5}{12}$

    (D) 12

    (E) $\frac{12}{5}$

3. SIMPLIFYING FRACTIONS

All fractional answers should be left in simplest form. There should be no factor that can be divided into numerator and denominator. In simplifying fractions involving very large numbers, it is helpful to tell at a glance whether or not a given number will divide evenly into both numerator and denominator. Certain tests for divisibility assist with this.

If a number is divisible by	Then
2	its last digit is 0, 2, 4, 6, or 8
3	the sum of the digits is divisible by 3
4	the number formed by the last 2 digits is divisible by 4
5	the last digit is 5 or 0
6	the number meets the tests for divisibility by 2 and 3
8	the number formed by the last 3 digits is divisible by 8
9	the sum of the digits is divisible by 9

Example:

By what single digit number should we simplify $\dfrac{135,492}{428,376}$?

Solution:

Since both numbers are even, they are at least divisible by 2. The sum of the digits in the numerator is 24. The sum of the digits in the denominator is 30. Since these sums are both divisible by 3, each number is divisible by 3. Since these numbers meet the divisibility tests for 2 and 3, they are each divisible by 6.

Example:

Simplify to simplest form: $\dfrac{43,672}{52,832}$

Solution:

Since both numbers are even, they are at least divisible by 2. However, to save time, we would like to divide by a larger number. The sum of the digits in the numerator is 22, so it is not divisible by 3. The number formed by the last two digits of each number is divisible by 4, making the entire number divisible by 4. The numbers formed by the last three digits of each number is divisible by 8. Therefore, each number is divisible by 8. Dividing by 8, we have $\dfrac{5459}{6604}$. Since these numbers are no longer even and divisibility by 3 was ruled out earlier, there is no longer a single digit factor common to numerator and denominator. It is unlikely, at the level of this examination, that you will be called on to divide by a two-digit number.

Exercise 3

Work out each problem. Circle the letter that appears before your answer.

1. Which of the following numbers is divisible by 5 and 9?

   (A) 42,235
   (B) 34,325
   (C) 46,505
   (D) 37,845
   (E) 53,290

2. Given the number 83,21p, in order for this number to be divisible by 3, 6, and 9, p must be

   (A) 4
   (B) 5
   (C) 6
   (D) 0
   (E) 9

3. If $n!$ means $n(n - 1)(n - 2) \ldots (4)(3)(2)(1)$, so that $4! = (4)(3)(2)(1) = 24$, then 19! is divisible by

   I. 17
   II. 54
   III. 100
   IV. 39
   (A) I and II only
   (B) I only
   (C) I and IV only
   (D) I, II, III, and IV
   (E) none of the above

4. The fraction $\dfrac{432}{801}$ can be simplified by dividing numerator and denominator by

   (A) 2
   (B) 4
   (C) 6
   (D) 8
   (E) 9

5. The number 6,862,140 is divisible by

   I. 3
   II. 4
   III. 5
   (A) I only
   (B) I and III only
   (C) II and III only
   (D) I, II, and III
   (E) III only

4. OPERATIONS WITH MIXED NUMBERS

If you're dividing and using a calculator, put the answer you got over 1 ex: instead of 2, answer is 1/2

To add or subtract mixed numbers, it is again important to find common denominators. If it is necessary to borrow in subtraction, you must borrow in terms of the common denominator.

Example:

$$23\frac{1}{3} - 6\frac{2}{5}$$

Solution:

$$23\frac{1}{3} = 23\frac{5}{15}$$

$$-6\frac{2}{5} = -6\frac{6}{15}$$

Since we cannot subtract $\frac{6}{15}$ from $\frac{5}{15}$, we borrow $\frac{15}{15}$ from 23 and rewrite our problem as

$$22\frac{20}{15}$$
$$-6\frac{6}{15}$$

In this form, subtraction is possible, giving us an answer of $16\frac{14}{15}$.

Example:

Add $17\frac{3}{4}$ to $43\frac{3}{5}$

Solution:

Again we first rename the fractions to have a common denominator. This time it will be 20.

$$17\frac{3}{4} = 17\frac{15}{20}$$

$$+43\frac{3}{5} = +43\frac{12}{20}$$

When adding, we get a sum of $60\frac{27}{20}$, which we change to $61\frac{7}{20}$.

To multiply or divide mixed numbers, always rename them as improper fractions first.

Example:

Multiply $3\frac{3}{5} \cdot 1\frac{1}{9} \cdot 2\frac{3}{4}$

Solution:

$$\frac{\overset{2}{\cancel{18}}}{\cancel{5}} \cdot \frac{\overset{2}{\cancel{10}}}{\cancel{9}} \cdot \frac{11}{\underset{2}{\cancel{4}}} = 11$$

Example:

Divide $3\frac{3}{4}$ by $5\frac{5}{8}$

Solution:

$$\frac{15}{4} \div \frac{45}{8} = \frac{\cancel{15}}{\cancel{4}} \cdot \frac{\cancel{8}^{2}}{\cancel{45}_{3}} = \frac{2}{3}$$

Exercise 4

Work out each problem. Circle the letter that appears before your answer.

1. Find the sum of $1\frac{1}{6}$, $2\frac{2}{3}$, and $3\frac{3}{4}$.

    (A) $7\frac{5}{12}$

    (B) $6\frac{6}{13}$

    (C) $7\frac{7}{12}$

    (D) $6\frac{1}{3}$

    (E) $7\frac{1}{12}$

2. Subtract $45\frac{5}{12}$ from 61.

    (A) $15\frac{7}{12}$

    (B) $15\frac{5}{12}$

    (C) $16\frac{7}{12}$

    (D) $16\frac{5}{12}$

    (E) $17\frac{5}{12}$

3. Find the product of $32\frac{1}{2}$ and $5\frac{1}{5}$.

    (A) 26

    (B) 13

    (C) 169

    (D) $160\frac{1}{10}$

    (E) $160\frac{2}{7}$

4. Divide $17\frac{1}{2}$ by 70.

    (A) $\frac{1}{4}$

    (B) 4

    (C) $\frac{1}{2}$

    (D) $4\frac{1}{2}$

    (E) $\frac{4}{9}$

5. Find $1\frac{3}{4} \cdot 12 \div 8\frac{2}{5}$.

    (A) $\frac{2}{5}$

    (B) $\frac{5}{288}$

    (C) $2\frac{1}{5}$

    (D) $\frac{1}{2}$

    (E) $2\frac{1}{2}$

5. COMPARING FRACTIONS

There are two methods by which fractions may be compared to see which is larger (or smaller).

Method I—Rename the fractions to have the same denominator. When this is done, the fraction with the larger numerator is the larger fraction.

Example:

Which is larger, $\dfrac{5}{6}$ or $\dfrac{8}{11}$?

Solution:

The least common denominator is 66.

$$\dfrac{5}{6} = \dfrac{55}{66} \qquad \dfrac{8}{11} = \dfrac{48}{66}$$

Therefore, $\dfrac{5}{6}$ is the larger fraction.

Method II—To compare $\dfrac{a}{b}$ with $\dfrac{c}{d}$, compare the cross products as follows:

If $ad > bc$, then $\dfrac{a}{b} > \dfrac{c}{d}$

If $ad < bc$, then $\dfrac{a}{b} < \dfrac{c}{d}$

If $ad = bc$, then $\dfrac{a}{b} = \dfrac{c}{d}$

Using the example above, to compare $\dfrac{5}{6}$ with $\dfrac{8}{11}$, compare $5 \cdot 11$ with $6 \cdot 8$. Since $5 \cdot 11$ is greater, $\dfrac{5}{6}$ is the larger fraction.

Sometimes, a combination of these methods must be used in comparing a series of fractions. When a common denominator can be found easily for a series of fractions, Method I is easier. When a common denominator would result in a very large number, Method II is easier.

Example:

Which of the following fractions is the largest?

(A) $\dfrac{3}{5}$

(B) $\dfrac{21}{32}$

(C) $\dfrac{11}{16}$

(D) $\dfrac{55}{64}$

(E) $\dfrac{7}{8}$

Solution:

To compare the last four, we can easily use a common denominator of 64.

$$\dfrac{21}{32} = \dfrac{42}{64} \qquad \dfrac{11}{16} = \dfrac{44}{64} \qquad \dfrac{55}{64} \qquad \dfrac{7}{8} = \dfrac{56}{64}$$

The largest of these is $\dfrac{7}{8}$. Now we compare $\dfrac{7}{8}$ with $\dfrac{3}{5}$ using Method II. $7 \cdot 5 > 8 \cdot 3$; therefore, $\dfrac{7}{8}$ is the greatest fraction.

Exercise 5

Work out each problem. Circle the letter that appears before your answer.

1. Arrange these fractions in order of size, from largest to smallest: $\frac{4}{15}, \frac{2}{5}, \frac{1}{3}$.

   (A) $\frac{4}{15}, \frac{2}{5}, \frac{1}{3}$

   (B) $\frac{4}{15}, \frac{1}{3}, \frac{2}{5}$

   (C) $\frac{2}{5}, \frac{1}{3}, \frac{4}{15}$

   (D) $\frac{1}{3}, \frac{4}{15}, \frac{2}{5}$

   (E) $\frac{1}{3}, \frac{2}{5}, \frac{4}{15}$

2. Which of the following fractions is the smallest?

   (A) $\frac{3}{4}$

   (B) $\frac{5}{6}$

   (C) $\frac{7}{8}$

   (D) $\frac{19}{24}$

   (E) $\frac{13}{15}$

3. Which of the following fractions is the largest?

   (A) $\frac{3}{5}$

   (B) $\frac{7}{10}$

   (C) $\frac{5}{8}$

   (D) $\frac{3}{4}$

   (E) $\frac{13}{20}$

4. Which of the following fractions is closest to $\frac{3}{4}$?

   (A) $\frac{1}{2}$

   (B) $\frac{7}{12}$

   (C) $\frac{5}{6}$

   (D) $\frac{11}{12}$

   (E) $\frac{19}{24}$

5. Which of the following fractions is closest to $\frac{1}{2}$?

   (A) $\frac{5}{12}$

   (B) $\frac{8}{15}$

   (C) $\frac{11}{20}$

   (D) $\frac{31}{60}$

   (E) $\frac{7}{15}$

6. COMPLEX FRACTIONS

To simplify complex fractions, fractions that contain fractions within them, multiply every term by the smallest number needed to clear all fractions in the given numerator and denominator.

Example:

$$\dfrac{\dfrac{1}{6}+\dfrac{1}{4}}{\dfrac{1}{2}+\dfrac{1}{3}}$$

Solution:

The smallest number into which 6, 4, 2, and 3 will divide is 12. Therefore, multiply every term of the fraction by 12 to simplify the fraction.

$$\dfrac{2+3}{6+4}=\dfrac{5}{10}=\dfrac{1}{2}$$

Example:

$$\dfrac{\dfrac{3}{4}-\dfrac{2}{3}}{1+\dfrac{1}{2}}$$

Solution:

Again, we multiply every term by 12. Be sure to multiply the 1 by 12 also.

$$\dfrac{9-8}{12+6}=\dfrac{1}{18}$$

Exercise 6

Work out each problem. Circle the letter that appears before your answer.

1. Write as a fraction in simplest form: $\dfrac{\dfrac{2}{3}+\dfrac{1}{6}+\dfrac{1}{4}}{\dfrac{2}{3}-\dfrac{1}{2}}$

(A) $\dfrac{13}{2}$

(B) $\dfrac{7}{2}$

(C) $\dfrac{13}{4}$

(D) $\dfrac{4}{13}$

(E) $\dfrac{49}{12}$

2. Simplify: $\dfrac{\dfrac{5}{6}-\dfrac{2}{3}}{\dfrac{5}{12}-\dfrac{1}{6}}$

(A) $\dfrac{5}{12}$

(B) $\dfrac{5}{6}$

(C) $\dfrac{2}{3}$

(D) $\dfrac{1}{6}$

(E) $\dfrac{7}{12}$

3. Find the value of $\dfrac{\dfrac{1}{a}+\dfrac{1}{b}}{\dfrac{1}{ab}}$ when $a = 2$ and $b = 3$.

    (A) $\dfrac{5}{6}$

    (B) 5

    (C) $4\dfrac{1}{6}$

    (D) $1\dfrac{1}{5}$

    (E) $2\dfrac{2}{5}$

4. Find the value of $\dfrac{\dfrac{1}{a}+\dfrac{1}{b}}{\dfrac{1}{ab}}$ when $a = \dfrac{1}{2}$ and $b = \dfrac{1}{3}$.

    (A) $\dfrac{5}{6}$

    (B) 5

    (C) $4\dfrac{1}{6}$

    (D) $1\dfrac{1}{5}$

    (E) $2\dfrac{2}{5}$

5. Find the value of $\dfrac{2\dfrac{1}{3}}{5\dfrac{1}{2}+3\dfrac{1}{3}}$.

    (A) $\dfrac{4}{17}$

    (B) $\dfrac{21}{25}$

    (C) $\dfrac{7}{6}$

    (D) $\dfrac{12}{51}$

    (E) $\dfrac{14}{53}$

RETEST

Work out each problem. Circle the letter that appears before your answer.

1. The sum of $\frac{4}{5}$, $\frac{3}{4}$, and $\frac{1}{3}$ is

   (A) $\frac{8}{12}$

   (B) $\frac{113}{60}$

   (C) $\frac{1}{5}$

   (D) $\frac{10}{9}$

   (E) $\frac{11}{6}$

2. Subtract $\frac{2}{3}$ from $\frac{11}{15}$.

   (A) $\frac{3}{4}$

   (B) $\frac{7}{5}$

   (C) $\frac{5}{7}$

   (D) $\frac{1}{15}$

   (E) $\frac{1}{3}$

3. If $52,34p$ is divisible by 9, the digit represented by p must be
   (A) 1
   (B) 2
   (C) 3
   (D) 4
   (E) 5

4. $\left(\frac{3}{5}+\frac{1}{4}\right)\div\frac{34}{15}$ is equal to

   (A) $\frac{5}{3}$

   (B) $\frac{5}{8}$

   (C) $\frac{8}{3}$

   (D) $\frac{8}{5}$

   (E) $\frac{3}{8}$

5. Subtract $62\frac{2}{3}$ from 100.

   (A) $37\frac{1}{3}$

   (B) $38\frac{1}{3}$

   (C) $37\frac{2}{3}$

   (D) $38\frac{2}{3}$

   (E) $28\frac{2}{3}$

6. Divide $2\frac{2}{5}$ by $4\frac{8}{10}$.

   (A) 2

   (B) $\frac{1}{2}$

   (C) $\frac{288}{25}$

   (D) $\frac{25}{288}$

   (E) $2\frac{1}{4}$

7. Which of the following fractions is the smallest?

   (A) $\frac{7}{12}$

   (B) $\frac{8}{15}$

   (C) $\frac{11}{20}$

   (D) $\frac{5}{6}$

   (E) $\frac{2}{3}$

8. Which of the following fractions is closest to $\frac{1}{4}$?

   (A) $\frac{4}{15}$

   (B) $\frac{3}{10}$

   (C) $\frac{3}{20}$

   (D) $\frac{1}{5}$

   (E) $\frac{1}{10}$

9. Simplify: $\dfrac{\dfrac{5}{2}+\dfrac{2}{3}}{\dfrac{3}{4}+\dfrac{5}{6}}$

    (A) 2

    (B) $\dfrac{1}{2}$

    (C) 12

    (D) $\dfrac{1}{4}$

    (E) 4

10. Find the value of $\dfrac{\dfrac{1}{ab}}{\dfrac{1}{a}+\dfrac{1}{b}}$ when $a = 4$, $b = 5$.

    (A) 9

    (B) 20

    (C) $\dfrac{1}{9}$

    (D) $\dfrac{1}{20}$

    (E) $\dfrac{9}{40}$

SOLUTIONS TO PRACTICE EXERCISES

Diagnostic Test

1. **(D)** Change all fractions to sixtieths.

$$\frac{36}{60}+\frac{40}{60}+\frac{15}{60}=\frac{91}{60}$$

2. **(A)** $\frac{9}{10}-\frac{3}{4}=\frac{36-30}{40}=\frac{6}{40}=\frac{3}{20}$

3. **(D)** The sum of the digits is 27, which is divisible by 9.

4. **(C)** $\frac{5}{6}\div\left(\frac{4}{3}\cdot\frac{5}{4}\right)=\frac{5}{6}\div\frac{5}{3}=\frac{\cancel{5}}{\cancel{6}_2}\cdot\frac{\cancel{3}}{\cancel{5}}=\frac{1}{2}$

5. **(A)** $57\ =56\frac{5}{5}$

$$\begin{array}{r}32\frac{3}{5}=32\frac{3}{5}\\ \hline 24\frac{2}{5}\end{array}$$

6. **(B)** $\frac{9}{2}\div\frac{9}{8}=\frac{\cancel{9}}{\cancel{2}}\cdot\frac{\overset{4}{\cancel{8}}}{\cancel{9}}=4$

7. **(E)** Use a common denominator of 32.

$$\frac{1}{2}=\frac{16}{32}\qquad\frac{11}{16}=\frac{22}{32}\qquad\frac{5}{8}=\frac{20}{32}\qquad\frac{21}{32}$$
$$\frac{3}{4}=\frac{24}{32}$$

Of these, $\frac{3}{4}$ is the largest.

8. **(B)** Use a common denominator of 30.

$$\frac{11}{15}=\frac{22}{30}\qquad\frac{7}{10}=\frac{21}{30}\qquad\frac{4}{5}=\frac{24}{30}$$
$$\frac{1}{2}=\frac{15}{30}\qquad\frac{5}{6}=\frac{25}{30}$$

Since $\frac{2}{3}=\frac{20}{30}$, the answer closest to $\frac{2}{3}$ is $\frac{7}{10}$.

9. **(B)** Multiply every term of the fraction by 30.

$$\frac{120-27}{20+15}=\frac{93}{35}$$

10. **(A)** $\dfrac{\dfrac{1}{3}+\dfrac{1}{4}}{\dfrac{1}{3}-\dfrac{1}{4}}$

Multiply every term by 12.

$$\frac{4+3}{4-3}=7$$

Exercise 1

1. **(B)** Change all fractions to twelfths.

$$\frac{6}{12}+\frac{8}{12}+\frac{9}{12}=\frac{23}{12}$$

2. **(A)** Use the cross product method.

$$\frac{5(15)+17(3)}{17(15)}=\frac{75+51}{255}=\frac{126}{255}$$

3. **(D)** $\frac{3}{4}+\frac{5}{6}=\frac{18+20}{24}=\frac{38}{24}=\frac{19}{12}$

$$\frac{1}{4}+\frac{2}{3}=\frac{3+8}{12}=\frac{11}{12}$$

$$\frac{19}{12}-\frac{11}{12}=\frac{8}{12}=\frac{2}{3}$$

4. **(B)** $\frac{9}{11}-\frac{3}{5}=\frac{45-33}{55}=\frac{12}{55}$

5. **(E)** $\frac{1}{4}+\frac{2}{3}=\frac{3+8}{12}=\frac{11}{12}$

$$\frac{11}{12}-\frac{5}{8}=\frac{88-60}{96}=\frac{28}{96}=\frac{7}{24}$$

Exercise 2

1. **(B)** $\frac{\cancel{3}}{\cancel{2}}\cdot\frac{\cancel{6}}{1}\cdot\frac{\overset{2}{\cancel{4}}}{\cancel{9}_3}\cdot\frac{1}{\cancel{12}_2}=\frac{1}{3}$

2. **(D)** $\frac{7}{\cancel{8}}\cdot\frac{2}{3}\cdot\frac{\cancel{8}}{1}=\frac{14}{3}$

3. **(A)** $\frac{3}{5}\div\frac{3}{20}$

$$\frac{\cancel{3}}{\cancel{5}}\cdot\frac{\overset{4}{\cancel{20}}}{\cancel{3}}=4$$

4. **(E)** $\frac{\cancel{2}}{3}\cdot\frac{7}{\cancel{12}_6}=\frac{7}{18}$

5. **(D)** $\frac{\cancel{5}}{1}\cdot\frac{12}{\cancel{5}}=12$

Exercise 3

1. (D) The digits must add to a number divisible by 9. All answers are divisible by 5. $3 + 7 + 8 + 4 + 5 = 27$, which is divisible by 9.

2. (A) The sum of the digits must be divisible by 9, and the digit must be even. $8 + 3 + 2 + 1 = 14$. Therefore, we choose (A) because $14 + 4 = 18$, which is divisible by 9.

3. (D) $19! = 19 \cdot 18 \cdot 17 \cdot 16 \ldots 3 \cdot 2 \cdot 1$. This is divisible by 17, since it contains a factor of 17. It is divisible by 54, since it contains factors of 9 and 6. It is divisible by 100, since it contains factors of 10, 5, and 2. It is divisible by 39, since it contains factors of 13 and 3.

4. (E) The sum of the digits in both the numerator and denominator are divisible by 9.

5. (D) The sum of the digits is 27, which is divisible by 3. The number formed by the last two digits is 40, which is divisible by 4. The number ends in 0 and is therefore divisible by 5.

Exercise 4

1. (C) $1\frac{1}{6} = 1\frac{2}{12}$

   $2\frac{2}{3} = 2\frac{8}{12}$

   $3\frac{3}{4} = 3\frac{9}{12}$

   $6\frac{19}{12} = 7\frac{7}{12}$

2. (A) $61 = 60\frac{12}{12}$

   $\underline{45\frac{5}{12}} = \underline{45\frac{5}{12}}$

   $15\frac{7}{12}$

3. (C) $\dfrac{\overset{13}{\cancel{65}}}{\cancel{2}} \cdot \dfrac{\overset{13}{\cancel{26}}}{\cancel{5}} = 169$

4. (A) $17\frac{1}{2} \div 70 = \frac{35}{2} \div 70 = \frac{\cancel{35}}{2} \cdot \frac{1}{\underset{2}{\cancel{70}}} = \frac{1}{4}$

5. (E) $\dfrac{\cancel{7}}{\cancel{4}} \cdot \dfrac{\overset{3}{\cancel{12}}}{1} \cdot \dfrac{5}{\underset{6}{\cancel{42}}} = \frac{5}{2} = 2\frac{1}{2}$

Exercise 5

1. (C) $\frac{2}{5} = \frac{6}{15}$ $\frac{1}{3} = \frac{5}{15}$

2. (A) To compare (A), (B), (C), and (D), use a common denominator of 24.

   $\frac{3}{4} = \frac{18}{24}$ $\frac{5}{6} = \frac{20}{24}$ $\frac{7}{8} = \frac{21}{24}$ $\frac{19}{24}$

   Of these, $\frac{3}{4}$ is the smallest. To compare $\frac{3}{4}$ with $\frac{13}{15}$, use cross products. Since $(3)(15) < (4)(14)$, $\frac{3}{4} < \frac{13}{15}$. Therefore, (A) is the smallest.

3. (D) To compare (A), (B), (D), and (E), use a common denominator of 20.

   $\frac{3}{5} = \frac{12}{20}$ $\frac{7}{10} = \frac{14}{20}$ $\frac{3}{4} = \frac{15}{20}$ $\frac{13}{20}$

   Of these, $\frac{3}{4}$ is the largest. To compare $\frac{3}{4}$ with $\frac{5}{8}$, use cross products. Since $(3)(8) > (4)(5)$, $\frac{3}{4}$ is the larger fraction.

4. (E) Use a common denominator of 24.

   $\frac{1}{2} = \frac{12}{24}$ $\frac{7}{12} = \frac{14}{24}$ $\frac{5}{6} = \frac{20}{24}$ $\frac{11}{12} = \frac{22}{24}$

   $\frac{19}{24}$

   Since $\frac{3}{4} = \frac{18}{24}$, the answer closest to $\frac{3}{4}$ is (E), $\frac{19}{24}$.

5. (D) Use a common denominator of 60.

   $\frac{5}{12} = \frac{25}{60}$ $\frac{8}{15} = \frac{32}{60}$ $\frac{11}{20} = \frac{33}{60}$ $\frac{31}{60}$

   $\frac{7}{15} = \frac{28}{60}$

   Since $\frac{1}{2} = \frac{30}{60}$, the answer closest to $\frac{1}{2}$ is (D), $\frac{31}{60}$.

Exercise 6

1. **(A)** Multiply every term of the fraction by 12.

$$\frac{8+2+3}{8-6}=\frac{13}{2}$$

2. **(C)** Multiply every term of the fraction by 12.

$$\frac{10-8}{5-2}=\frac{2}{3}$$

3. **(B)** $\dfrac{\frac{1}{2}+\frac{1}{3}}{\frac{1}{6}}$ Multiply every term by 6.

$$\frac{3+2}{1}=5$$

4. **(A)** $\dfrac{1}{\frac{1}{2}}=2$ $\dfrac{1}{\frac{1}{3}}=3$ $\dfrac{1}{\frac{1}{6}}=6$

$$\frac{2+3}{6}=\frac{5}{6}$$

5. **(E)** $\dfrac{\frac{7}{3}}{\frac{11}{2}+\frac{10}{3}}$ Multiply every term by 6.

$$\frac{14}{33+20}=\frac{14}{53}$$

Retest

1. **(B)** Rename all fractions as sixtieths.

$$\frac{48}{60}+\frac{45}{60}+\frac{20}{60}=\frac{113}{60}$$

2. **(D)** $\dfrac{11}{15}-\dfrac{2}{3}=\dfrac{11}{15}-\dfrac{10}{15}=\dfrac{1}{15}$

3. **(D)** The sum of the digits must be divisible by 9.
$5 + 2 + 3 + 4 + 4 = 18$, which is divisible by 9.

4. **(E)** $\dfrac{17}{20}\div\dfrac{34}{15}$

$$\frac{\overset{1}{\cancel{17}}}{\underset{4}{\cancel{20}}}\cdot\frac{\overset{3}{\cancel{15}}}{\underset{2}{\cancel{34}}}=\frac{3}{8}$$

5. **(A)** $100\ \ =99\dfrac{3}{3}$

$$\dfrac{-62\dfrac{2}{3}=62\dfrac{2}{3}}{37\dfrac{1}{3}}$$

6. **(B)** $\dfrac{12}{5}\div\dfrac{48}{10}=\dfrac{\overset{1}{\cancel{12}}}{\cancel{5}}\cdot\dfrac{\overset{2}{\cancel{10}}}{\underset{4}{\cancel{48}}}=\dfrac{2}{4}=\dfrac{1}{2}$

7. **(B)** Use a common denominator of 60.

$$\frac{7}{12}=\frac{35}{60}\qquad\frac{8}{15}=\frac{32}{60}\qquad\frac{11}{20}=\frac{33}{60}\qquad\frac{5}{6}=\frac{50}{60}$$

$$\frac{2}{3}=\frac{40}{60}$$

Of these, $\dfrac{8}{15}$ is the smallest.

8. **(A)** Use a common denominator of 60.

$$\frac{4}{15}=\frac{16}{60}\qquad\frac{3}{10}=\frac{18}{60}\qquad\frac{3}{20}=\frac{9}{60}\qquad\frac{1}{5}=\frac{12}{60}$$

$$\frac{1}{10}=\frac{6}{60}$$

Since $\dfrac{1}{4}=\dfrac{15}{60}$, the answer closest to $\dfrac{1}{4}$ is $\dfrac{4}{15}$.

9. **(A)** Multiply every term of the fraction by 12.

$$\frac{30+8}{9+10}=\frac{38}{19}=2$$

10. **(C)** $\dfrac{\frac{1}{20}}{\frac{1}{4}+\frac{1}{5}}$ Multiply every term by 20.

$$\frac{1}{5+4}=\frac{1}{9}$$

Verbal Problems Involving Fractions

DIAGNOSTIC TEST

Directions: Work out each problem. Circle the letter that appears before your answer.

Answers are at the end of the chapter.

1. On Monday evening, Channel 2 scheduled 2 hours of situation comedy, 1 hour of news, and 3 hours of movies. What part of the evening's programming was devoted to situation comedy?

    (A) $\frac{1}{3}$

    (B) $\frac{2}{3}$

    (C) $\frac{1}{2}$

    (D) $\frac{1}{6}$

    (E) $\frac{2}{5}$

2. What part of a gallon is 2 qt. 1 pt.?

    (A) $\frac{3}{4}$

    (B) $\frac{3}{10}$

    (C) $\frac{1}{2}$

    (D) $\frac{5}{8}$

    (E) $\frac{3}{8}$

3. Michelle spent $\frac{1}{2}$ of her summer vacation at camp, $\frac{1}{5}$ of her vacation babysitting, and $\frac{1}{4}$ visiting her grandmother. What part of her vacation was left to relax at home?

    (A) $\frac{1}{5}$

    (B) $\frac{1}{20}$

    (C) $\frac{1}{3}$

    (D) $\frac{3}{20}$

    (E) $\frac{1}{6}$

4. After doing $\frac{1}{3}$ of the family laundry before breakfast, Mrs. Strauss did $\frac{3}{4}$ of the remainder before lunch. What part of the laundry was left for the afternoon?

    (A) $\frac{1}{2}$

    (B) $\frac{1}{4}$

    (C) $\frac{2}{3}$

    (D) $\frac{1}{5}$

    (E) $\frac{1}{6}$

5. Glenn spent $\frac{2}{5}$ of his allowance on a hit record. He then spent $\frac{2}{3}$ of the remainder on a gift. What part of his allowance did he have left?

    (A) $\frac{1}{5}$

    (B) $\frac{1}{3}$

    (C) $\frac{2}{5}$

    (D) $\frac{3}{20}$

    (E) $\frac{1}{10}$

6. Barbara's car has a gasoline tank that holds 20 gallons. When her gauge reads $\frac{1}{4}$ full, how many gallons are needed to fill the tank?

    (A) 5
    (B) 10
    (C) 15
    (D) 12
    (E) 16

7. 42 seniors voted to hold the prom at the Copacabana. This represents $\frac{2}{9}$ of the senior class. How many seniors did not vote for the Copacabana?

    (A) 147
    (B) 101
    (C) 189
    (D) 105
    (E) 126

8. Steve needs M hours to mow the lawn. After working for X hours, what part of the job remains to be done?

    (A) $\frac{M-X}{M}$

    (B) $M - \frac{X}{M}$

    (C) $M - X$

    (D) $X - M$

    (E) $\frac{X}{M}$

9. Of D dogs in Mrs. Pace's kennel, $\frac{1}{3}$ are classified as large dogs and $\frac{1}{4}$ of the remainder are classified as medium-sized. How many of the dogs are classified as small?

    (A) $\frac{1}{2}D$

    (B) $\frac{1}{6}D$

    (C) $\frac{5}{6}D$

    (D) $\frac{2}{3}D$

    (E) $\frac{1}{3}D$

10. A bookshelf contains A autobiographies and B biographies. What part of these books are biographies?

    (A) $\frac{B}{A}$

    (B) $\frac{B}{A+B}$

    (C) $\frac{A}{A}+B$

    (D) $\frac{A}{B}$

    (E) $\frac{B}{A}-B$

1. PART OF A WHOLE

A fraction represents a part of a whole. In dealing with fractional problems, we are usually dealing with a part of a quantity.

Example:

Andrea and Danny ran for president of the Math Club. Andrea got 15 votes, while Danny got the other 10. What part of the votes did Andrea receive?

Solution:

Andrea got 15 votes out of 25. That is $\frac{15}{25}$ or $\frac{3}{5}$ of the votes.

Exercise 1

Work out each problem. Circle the letter that appears before your answer.

1. In a class there are 18 boys and 12 girls. What part of the class is girls?

   (A) $\frac{2}{3}$

   (B) $\frac{3}{5}$

   (C) $\frac{2}{5}$

   (D) $\frac{1}{15}$

   (E) $\frac{3}{2}$

2. A team played 40 games and lost 6. What part of the games played did it win?

   (A) $\frac{3}{20}$

   (B) $\frac{3}{17}$

   (C) $\frac{14}{17}$

   (D) $\frac{17}{20}$

   (E) $\frac{7}{8}$

3. What part of an hour elapses between 3:45 p.m. and 4:09 p.m.?

   (A) $\frac{6}{25}$

   (B) $\frac{2}{5}$

   (C) $\frac{5}{12}$

   (D) $\frac{1}{24}$

   (E) 24

4. A camp employs 4 men, 6 women, 12 girls, and 8 boys. In the middle of the summer, 3 girls are fired and replaced by women. What part of the staff is then made up of women?

   (A) $\frac{1}{5}$

   (B) $\frac{2}{9}$

   (C) $\frac{1}{3}$

   (D) $\frac{3}{10}$

   (E) $\frac{1}{2}$

5. There are three times as many seniors as juniors at a high school Junior-Senior dance. What part of the students present are juniors?

   (A) $\dfrac{2}{5}$

   (B) $\dfrac{3}{5}$

   (C) $\dfrac{2}{3}$

   (D) $\dfrac{3}{4}$

   (E) $\dfrac{1}{4}$

6. What part of a yard is 1 ft. 3 in.?

   (A) $\dfrac{5}{12}$

   (B) $\dfrac{1}{3}$

   (C) $\dfrac{1}{2}$

   (D) $\dfrac{5}{8}$

   (E) $\dfrac{4}{9}$

7. Manorville High had a meeting of the Student Senate, which was attended by 10 freshmen, 8 sophomores, 15 juniors, and 7 seniors. What part of the students present at the meeting were sophomores?

   (A) $\dfrac{1}{4}$

   (B) $\dfrac{5}{8}$

   (C) $\dfrac{7}{40}$

   (D) $\dfrac{1}{5}$

   (E) $\dfrac{1}{3}$

8. The Dobkin family budgets its monthly income as follows: $\dfrac{1}{3}$ for food, $\dfrac{1}{4}$ for rent, $\dfrac{1}{10}$ for clothing, and $\dfrac{1}{5}$ for savings. What part is left for other expenses?

   (A) $\dfrac{3}{7}$

   (B) $\dfrac{1}{6}$

   (C) $\dfrac{7}{60}$

   (D) $\dfrac{2}{15}$

   (E) $\dfrac{3}{20}$

2. FINDING FRACTIONS OF FRACTIONS

Many problems require you to find a fractional part of a fractional part, such as $\frac{3}{5}$ of $\frac{2}{3}$. This involves multiplying the fractions together, $\frac{3}{4}$ of $\frac{2}{3}$ is $\frac{1}{2}$.

Example:

$\frac{1}{4}$ of the employees of Mr. Brown's firm earn over $20,000 per year. $\frac{1}{2}$ of the remainder earn between $15,000 and $20,000. What part of the employees earns less than $15,000 per year?

Solution:

$\frac{1}{4}$ earn over $20,000. $\frac{1}{2}$ of $\frac{3}{4}$ or $\frac{3}{8}$ earn between $15,000 and $20,000. That accounts for $\frac{1}{4} + \frac{3}{8}$ or $\frac{5}{8}$ of all employees. Therefore, the other $\frac{3}{8}$ earn less than $15,000.

Example:

A full bottle of isopropyl alcohol is left open in the school laboratory. If $\frac{1}{3}$ of the isopropyl alcohol evaporates in the first 12 hours and $\frac{2}{3}$ of the remainder evaporates in the second 12 hours, what part of the bottle is full at the end of 24 hours?

Solution:

$\frac{1}{3}$ evaporates during the first 12 hours. $\frac{2}{3}$ of $\frac{2}{3}$ or $\frac{4}{9}$ evaporates during the second 12 hours. This accounts for $\frac{7}{9}$ of the isopropyl alcohol. Therefore, $\frac{2}{9}$ of the bottle is still full.

Exercise 2

Work out each problem. Circle the letter that appears before your answer.

1. Mrs. Natt spent $\frac{2}{3}$ of the family income one year and divided the remainder between 4 different savings banks. If she put $2000 into each bank, what was the amount of her family income that year?

    (A) $8000
    (B) $16,000
    (C) $24,000
    (D) $32,000
    (E) $6000

2. After selling $\frac{2}{5}$ of the suits in his shop before Christmas, Mr. Gross sold the remainder of the suits at the same price per suit after Christmas for $4500. What was the income from the entire stock?

    (A) $3000
    (B) $7500
    (C) $1800
    (D) $2700
    (E) $8000

3. Of this year's graduating seniors at South High, $\frac{9}{10}$ will be going to college. Of these, $\frac{4}{5}$ will go to four-year colleges, while the rest will be going to two-year colleges. What part of the class will be going to two-year colleges?

   (A) $\frac{9}{50}$

   (B) $\frac{1}{5}$

   (C) $\frac{4}{5}$

   (D) $\frac{18}{25}$

   (E) $\frac{4}{25}$

4. Sue and Judy drove from New York to San Francisco, a distance of 3000 miles. They covered $\frac{1}{10}$ of the distance the first day and $\frac{2}{9}$ of the remaining distance the second day. How many miles were left to be driven?

   (A) 600
   (B) 2000
   (C) 2400
   (D) 2100
   (E) 2700

5. 800 employees work for the Metropolitan Transportation Company. $\frac{1}{4}$ of these are college graduates, while $\frac{5}{6}$ of the remainder are high school graduates. What part of the employees never graduated from high school?

   (A) $\frac{1}{6}$

   (B) $\frac{1}{8}$

   (C) $\frac{7}{8}$

   (D) $\frac{1}{12}$

   (E) $\frac{3}{4}$

3. FINDING WHOLE NUMBERS

When a fractional part of a number is given and we wish to find the number representing the whole, it is often easiest to translate the words into mathematical symbols and solve the resulting equation.

Example:

Norman buys a used car for $2400, which is $\frac{2}{5}$ of the original price. Find the original price.

Solution:

$$2400 = \frac{2}{5}x \quad \text{Multiply by 5.}$$
$$12000 = 2x$$
$$\$6000 = x$$

Example:

The gas gauge on Mary's car reads $\frac{1}{8}$ full. She asks the gasoline attendant to fill the tank and finds she needs 21 gallons. What is the capacity of her gas tank?

Solution:

$\frac{7}{8}$ of the tank is empty and requires 21 gallons to fill.

$$\frac{7}{8}x = 21 \quad \text{Multiply by 8.}$$
$$7x = 168$$
$$x = 24$$

Exercise 3

Work out each problem. Circle the letter that appears before your answer.

1. Daniel spent $4.50 for a ticket to the movies. This represents $\frac{3}{4}$ of his allowance for the week. What did he have left that week for other expenses?

    (A) $6.00
    (B) $4.00
    (C) $3.39
    (D) $1.13
    (E) $1.50

2. 350 seniors attended the prom. This represents $\frac{7}{9}$ of the class. How many seniors did not attend the prom?

    (A) 50
    (B) 100
    (C) 110
    (D) 120
    (E) 450

3. A resolution was passed by a ratio of 5:4. If 900 people voted for the resolution, how many voted against it?

    (A) 500
    (B) 400
    (C) 720
    (D) 600
    (E) 223

4. Mr. Rich owns $\frac{2}{7}$ of a piece of property. If the value of his share is $14,000, what is the total value of the property?

    (A) $70,000
    (B) $49,000
    (C) $98,000
    (D) $10,000
    (E) $35,000

5. The Stone family spends $500 per month for rent. This is $\frac{4}{15}$ of their total monthly income. Assuming that salaries remain constant, what is the Stone family income for one year?

    (A) $1875
    (B) $6000
    (C) $60,000
    (D) $22,500
    (E) $16,000

4. SOLVING WITH LETTERS

When problems use letters in place of numbers, the same principles discussed earlier apply. If you are not sure which operations to use, replace the letters with numbers to determine the steps needed in the solution.

Example:

It takes Mr. Cohen X days to paint his house. If he works for D days, what part of his house must still be painted?

Solution:

He has $X - D$ days of painting left to do out of a total of X days; therefore, $\dfrac{X - D}{X}$ is the correct answer.

Example:

Sue buys 500 stamps. X of these are 10-cent stamps. $\dfrac{1}{3}$ of the remainder are 15-cent stamps. How many 15-cent stamps does she buy?

Solution:

She buys $500 - X$ stamps that are not 10-cents stamps. $\dfrac{1}{3}$ of these are 15-cent stamps. Therefore, she buys $\dfrac{1}{3}(500 - X)$ or $\dfrac{500 - X}{3}$ 15-cent stamps.

Example:

John spent $\$X$ on the latest hit record album. This represents $\dfrac{1}{M}$ of his weekly allowance. What is his weekly allowance?

Solution:

Translate the sentence into an algebraic equation.

Let A = weekly allowance

$X = \dfrac{1}{M} \cdot A$ Multiply by M.

$MX = A$

Exercise 4

Work out each problem. Circle the letter that appears before your answer.

1. A class contains B boys and G girls. What part of the class is boys?

   (A) $\dfrac{B}{G}$

   (B) $\dfrac{G}{B}$

   (C) $\dfrac{B}{B+G}$

   (D) $\dfrac{B+G}{B}$

   (E) $\dfrac{B}{B-G}$

2. M men agreed to rent a ski lodge for a total of D dollars. By the time they signed the contract, the price had increased by $100. Find the amount each man had to contribute as his total share.

   (A) $\dfrac{D}{M}$

   (B) $\dfrac{D}{M}+100$

   (C) $\dfrac{D+100}{M}$

   (D) $\dfrac{M}{D}+100$

   (E) $\dfrac{M+100}{D}$

3. Of S students in Bryant High, $\dfrac{1}{3}$ study French. $\dfrac{1}{4}$ of the remainder study Italian. How many of the students study Italian?

   (A) $\dfrac{1}{6}S$

   (B) $\dfrac{1}{4}S$

   (C) $\dfrac{2}{3}S$

   (D) $\dfrac{1}{12}S$

   (E) $\dfrac{3}{7}S$

4. Mr. and Mrs. Feldman took t dollars in travelers checks with them on a trip. During the first week, they spent $\dfrac{1}{5}$ of their money. During the second week, they spent $\dfrac{1}{3}$ of the remainder. How much did they have left at the end of the second week?

   (A) $\dfrac{4t}{15}$

   (B) $\dfrac{t}{15}$

   (C) $\dfrac{7t}{15}$

   (D) $\dfrac{11t}{15}$

   (E) $\dfrac{8t}{15}$

5. Frank's gas tank was $\dfrac{1}{4}$ full. After putting in G gallons of gasoline, the tank was $\dfrac{7}{8}$ full. What was the capacity of the tank.

   (A) $\dfrac{5G}{8}$

   (B) $\dfrac{8G}{5}$

   (C) $\dfrac{8G}{7}$

   (D) $\dfrac{7G}{8}$

   (E) $4G$

RETEST

Work out each problem. Circle the letter that appears before your answer.

1. The All Star Appliance Shop sold 10 refrigerators, 8 ranges, 12 freezers, 12 washing machines, and 8 clothes dryers during January. Freezers made up what part of the appliances sold in January?

    (A) $\dfrac{12}{50}$

    (B) $\dfrac{12}{25}$

    (C) $\dfrac{1}{2}$

    (D) $\dfrac{12}{40}$

    (E) $\dfrac{12}{60}$

2. What part of a day is 4 hours 20 minutes?

    (A) $\dfrac{1}{6}$

    (B) $\dfrac{13}{300}$

    (C) $\dfrac{1}{3}$

    (D) $\dfrac{13}{72}$

    (E) $\dfrac{15}{77}$

3. Mrs. Brown owns X books. $\dfrac{1}{3}$ of these are novels, $\dfrac{2}{5}$ of the remainder are poetry, and the rest are nonfiction. How many nonfiction books does Mrs. Brown own?

    (A) $\dfrac{4}{15}X$

    (B) $\dfrac{2}{5}X$

    (C) $\dfrac{2}{3}X$

    (D) $\dfrac{3}{5}X$

    (E) $\dfrac{7}{15}X$

4. After typing $\dfrac{1}{4}$ of a term paper on Friday, Richard completed $\dfrac{2}{3}$ of the remainder on Saturday. If he wanted to finish the paper that weekend, what part was left to be typed on Sunday?

    (A) $\dfrac{1}{4}$

    (B) $\dfrac{2}{3}$

    (C) $\dfrac{1}{3}$

    (D) $\dfrac{1}{2}$

    (E) $\dfrac{5}{6}$

5. What part of an hour elapses between 6:51 P.M. and 7:27 P.M.?

    (A) $\dfrac{1}{2}$

    (B) $\dfrac{2}{3}$

    (C) $\dfrac{3}{5}$

    (D) $\dfrac{17}{30}$

    (E) $\dfrac{7}{12}$

6. Laurie spent 8 hours reading a novel. If she finished $\dfrac{2}{5}$ of the book, how many more hours will she need to read the rest of the book?

    (A) 20

    (B) 12

    (C) $3\dfrac{1}{5}$

    (D) 18

    (E) 10

7. Mrs. Bach spent $\frac{2}{7}$ of her weekly grocery money on produce. If she spent $28 on produce, what was her total grocery bill that week?

   (A) $70
   (B) $80
   (C) $56
   (D) $90
   (E) $98

8. After working on a new roof for X hours on Saturday, Mr. Goldman finished the job by working Y hours on Sunday. What part of the total job was done on Sunday?

   (A) $\dfrac{Y}{X+Y}$

   (B) $\dfrac{Y}{X}$

   (C) $\dfrac{X}{X+Y}$

   (D) $\dfrac{Y}{X-Y}$

   (E) $\dfrac{Y}{Y-X}$

9. $\frac{1}{2}$ of the women in the Spring Garden Club are over 60 years old. $\frac{1}{4}$ of the remainder are under 40. What part of the membership is between 40 and 60 years old?

   (A) $\dfrac{1}{4}$

   (B) $\dfrac{3}{8}$

   (C) $\dfrac{3}{4}$

   (D) $\dfrac{1}{8}$

   (E) $\dfrac{5}{8}$

10. A residential city block contains R one-family homes, S two-family homes, and T apartment houses. What part of the buildings on this block is made up of one or two family houses?

    (A) $\dfrac{R}{T}+\dfrac{S}{T}$

    (B) $\dfrac{RS}{R+S+T}$

    (C) $\dfrac{R+S}{R+S+T}$

    (D) $\dfrac{R+S}{RST}$

    (E) $R+S$

SOLUTIONS TO PRACTICE EXERCISES

Diagnostic Test

1. (A) There was a total of 6 hours of programming time. $\frac{2}{6} = \frac{1}{3}$

2. (D) Change all measurements to pints. One gallon is 8 pints. 2 qt. 1 pt. = 5 pints = $\frac{5}{8}$ gallon.

3. (B) $\frac{1}{2} + \frac{1}{5} + \frac{1}{4} = \frac{10}{20} + \frac{4}{20} + \frac{5}{20} = \frac{19}{20}$. Therefore, $\frac{1}{20}$ was left to relax.

4. (E) $\frac{3}{4}$ of $\frac{2}{3}$ or $\frac{1}{2}$ of the laundry was done before lunch. Since $\frac{1}{3}$ was done before breakfast, $\frac{1}{3} + \frac{1}{2}$ or $\frac{5}{6}$ was done before the afternoon, leaving $\frac{1}{6}$ for the afternoon.

5. (A) $\frac{2}{3}$ of $\frac{3}{5}$ or $\frac{2}{5}$ of Glenn's allowance was spent on a gift. Since $\frac{2}{5}$ was spent on a hit record, $\frac{2}{5} + \frac{2}{5}$ or $\frac{4}{5}$ was spent, leaving $\frac{1}{5}$.

6. (C) The tank contained $\frac{1}{4} \cdot 20$ or 5 gallons, leaving 15 gallons to fill the tank.

7. (C) $42 = \frac{2}{9}x$ Multiply by 9. Divide by 2.
   $378 = 2x$
   $189 = x$

   This is the number of seniors. Since 42 seniors voted for the Copacabana, 147 did not.

8. (A) After working for X hours, $M - X$ hours are left out of a total of M hours.

9. (A) $\frac{1}{3}D$ dogs are large. $\frac{1}{4}$ of $\frac{2}{3}D$ or $\frac{1}{6}D$ are medium. The total of these dogs is $\frac{1}{3}D + \frac{1}{6}D$, leaving $\frac{1}{2}D$ small dogs.

10. (B) There are $A + B$ books. B out of $A + B$ are biographies.

Exercise 1

1. (C) There are 30 pupils in the class, of which 12 are girls. Therefore, $\frac{12}{30}$ or $\frac{2}{5}$ of the class is made up of girls.

2. (D) The team won 34 games out of 40 or $\frac{34}{40}$ of its games. This simplifies to $\frac{17}{20}$.

3. (B) 24 minutes is $\frac{24}{60}$ or $\frac{2}{5}$ of an hour.

4. (D) The number of staff members is still 30. Of these, 9 are now women. Therefore $\frac{9}{30}$ or $\frac{3}{10}$ of the staff are women.

5. (E) Let x = the number of juniors at the dance. $3x$ = the number of seniors at the dance. Then $4x$ = the number of students at the dance. x out of these $4x$ are juniors.

   That is $\frac{x}{4x}$ or $\frac{1}{4}$ of the students present are juniors.

6. (A) Change all measurements to inches. One yard is 36 inches. 1 ft. 3 in. is 15 inches.
   $\frac{15}{36} = \frac{5}{12}$

7. (D) There were 40 students at the meeting.
   $\frac{8}{40} = \frac{1}{5}$

8. (C) $\frac{1}{3} + \frac{1}{4} + \frac{1}{10} + \frac{1}{5} = \frac{20}{60} + \frac{15}{60} + \frac{6}{60} + \frac{12}{60} = \frac{53}{60}$
   Therefore, $\frac{7}{60}$ is left for other expenses.

Exercise 2

1. (C) She put $8000 into savings banks.

$$800 = \frac{1}{3}x \qquad \text{Multiply by 3.}$$
$$\$24,000 = x$$

2. (B) $4500 = \frac{3}{5}x$ Multiply by $\frac{5}{3}$.
$$\$7500 = x$$

3. (A) Since $\frac{4}{5}$ of $\frac{9}{10}$ will go to four-year colleges, $\frac{1}{5}$ of $\frac{9}{10}$ or $\frac{9}{50}$ will go to two-year colleges.

4. (D) They covered $\frac{1}{10} \cdot 3000$ or 300 miles the first day, leaving 2700 miles still to drive. They covered $\frac{2}{9} \cdot 2700$ or 600 miles the second day, leaving 2100 miles still to drive.

5. (B) $\frac{5}{6}$ of $\frac{3}{4}$ or $\frac{5}{8}$ are high school graduates. Since $\frac{1}{4}$ are college graduates, $\frac{1}{4} + \frac{5}{8}$ or $\frac{7}{8}$ of the employees graduated from high school, leaving $\frac{1}{8}$ who did not.

Exercise 3

1. (E) $4.50 = \frac{3}{4}x$ Multiply by 4. Divide by 3.
$$18.00 = 3x$$

$x = \$6.00$, his allowance for the week. $6.00 - \$4.50 = \$1.50$ left for other expenses.

2. (B) $350 = \frac{7}{9}x$ Multiply by 9. Divide by 7.
$$3150 = 7x$$
$$450 = x$$

This is the number of students in the class. If 350 attend the prom, 100 do not.

3. (C) $\frac{5}{9}$ of the voters voted for the resolution.

$$900 = \frac{5}{9}x \quad \text{Multiply by 9. Divide by 5.}$$
$$8100 = 5x$$
$$1620 = x$$

1620 - 900 = 720 voted against the resolution.

4. (B) $\frac{2}{7}x = 14,000$ Multiply by 7. Divide by 2.
$$2x = 98,000$$
$$x = \$49,000$$

5. (D) $\frac{4}{15}x = 500$ Multiply by 15. Divide by 4. This is their *monthly* income.
$$4x = 7500$$
$$x = \$1875$$

Multiply by 12 to find yearly income: $22,500.

Exercise 4

1. (C) There are $B + G$ students in the class. B out of $B + G$ are boys.

2. (C) The total cost is $D + 100$, which must be divided by the number of men to find each share. Since there are M men, each man must contribute $\dfrac{D+100}{M}$ dollars.

3. (A) $\dfrac{1}{3}S$ students study French. $\dfrac{1}{4}$ of $\dfrac{2}{3}S$ or $\dfrac{1}{6}S$ study Italian.

4. (E) They spent $\dfrac{1}{5}t$ the first week. They spent $\dfrac{1}{3}$ of $\dfrac{4}{5}t$ or $\dfrac{4}{15}t$ the second week. During these two weeks they spent a total of $\dfrac{1}{5}t + \dfrac{4}{15}t$ or $\dfrac{7}{15}t$, leaving $\dfrac{8}{15}t$.

5. (B) The G gallons fill $\dfrac{7}{8} - \dfrac{1}{4}$ or $\dfrac{5}{8}$ of the tank.

   $\dfrac{5}{8}x = G$ \qquad Multiply by $\dfrac{8}{5}$.

   $x = \dfrac{8G}{5}$

Retest

1. (A) There were 50 appliances sold in January; $\dfrac{12}{50}$ were freezers.

2. (D) Change all measurements to minutes. One day is $60 \cdot 24$ or 1440 minutes. 4 hr. 20 min. = 260 min. $\dfrac{260}{1440} = \dfrac{13}{72}$

3. (B) $\dfrac{1}{3}X$ books are novels. $\dfrac{2}{5}$ of $\dfrac{2}{3}X$ or $\dfrac{4}{15}X$ are poetry. The total of these books is $\dfrac{1}{3}X + \dfrac{4}{15}X$ or $\dfrac{9}{15}X$, leaving $\dfrac{6}{15}X$ or $\dfrac{2}{5}X$ books which are nonfiction.

4. (A) $\dfrac{2}{3}$ of $\dfrac{3}{4}$ or $\dfrac{1}{2}$ of the term paper was completed on Saturday. Since $\dfrac{1}{4}$ was completed on Friday, $\dfrac{1}{4} + \dfrac{1}{2}$ or $\dfrac{3}{4}$ was completed before Sunday, leaving $\dfrac{1}{4}$ to be typed on Sunday.

5. (C) 36 minutes is $\dfrac{36}{60}$ or $\dfrac{3}{5}$ of an hour.

6. (B) $8 = \dfrac{2}{5}x$ Multiply by 5. Divide by 2.

   $40 = 2x$

   $20 = x$

   This is the total number of hours needed to read the book. Since Laurie already read for 8 hours, she will need 12 more hours to finish the book.

7. (E) $\dfrac{2}{7}x = 28$ \qquad Multiply by 7. Divide by 2.

   $2x = 196$

   $x = \$98$

8. (A) Mr. Goldman worked a total of $X + Y$ hours. Y out of $X + Y$ was done on Sunday.

9. (B) $\dfrac{1}{4}$ of $\dfrac{1}{2}$ or $\dfrac{1}{8}$ are under 40. Since $\dfrac{1}{2} + \dfrac{1}{8}$ or $\dfrac{5}{8}$ are over 60 or under 40, $\dfrac{3}{8}$ are between 40 and 60.

10. (C) There is a total of $R + S + T$ buildings on the block. $R + S$ out of $R + S + T$ are one or two family houses.

Variation

4

DIAGNOSTIC TEST

Directions: Work out each problem. Circle the letter that appears before your answer.

Answers are at the end of the chapter.

1. Solve for x: $\dfrac{2x}{3} = \dfrac{x+5}{4}$

   (A) 2
   (B) 3
   (C) 4
   (D) $4\frac{1}{2}$
   (E) 5

2. Solve for x if $a = 7$, $b = 8$, $c = 5$: $\dfrac{a-3}{x} = \dfrac{b+2}{4c}$

   (A) 4
   (B) 5
   (C) 6
   (D) 7
   (E) 8

3. A map is drawn using a scale of 2 inches = 25 miles. How far apart in miles are two cities which are $5\frac{2}{5}$ inches apart on the map?

   (A) 60
   (B) 65
   (C) $67\frac{1}{2}$
   (D) 69
   (E) 70

4. How many apples can be bought for c cents if n apples cost d cents?

   (A) $\dfrac{nc}{d}$
   (B) $\dfrac{nd}{c}$
   (C) $\dfrac{cd}{n}$
   (D) $\dfrac{d}{c}$
   (E) nc

5. Ms. Dehn drove 7000 miles during the first 5 months of the year. At this rate, how many miles will she drive in a full year?

   (A) 16,000
   (B) 16,800
   (C) 14,800
   (D) 15,000
   (E) 16,400

6. A gear having 20 teeth turns at 30 revolutions per minute and is meshed with another gear having 25 teeth. At how many revolutions per minute is the second gear turning?

   (A) 35
   (B) $37\frac{1}{2}$
   (C) $22\frac{1}{2}$
   (D) 30
   (E) 24

7. A boy weighing 90 pounds sits 3 feet from the fulcrum of a seesaw. His younger brother weighs 50 pounds. How far on the other side of the fulcrum should he sit to balance the seesaw?

   (A) $5\frac{3}{4}$ ft.
   (B) $5\frac{2}{5}$ ft.
   (C) $1\frac{2}{3}$ ft.
   (D) $1\frac{1}{3}$ ft.
   (E) $4\frac{1}{2}$ ft.

8. Alan has enough dog food to last his two dogs for three weeks. If a neighbor asks him to feed her dog as well, how long will the dog food last, assuming that all three dogs eat the same amount?

    (A) 10 days
    (B) 12 days
    (C) 14 days
    (D) 16 days
    (E) 18 days

9. A newspaper can be printed by m machines in h hours. If 2 of the machines are not working, how many hours will it take to print the paper?

    (A) $\dfrac{mh - 2h}{m}$

    (B) $\dfrac{m - 2}{mh}$

    (C) $\dfrac{mh + 2h}{m}$

    (D) $\dfrac{mh}{m - 2}$

    (E) $\dfrac{mh}{m + 2}$

10. An army platoon has enough rations to last 20 men for 6 days. If 4 more men join the group, for how many fewer days will the rations last?

    (A) 5
    (B) 2
    (C) 1
    (D) 1.8
    (E) 4

1. RATIO AND PROPORTION

A ratio is a comparison between two quantities. In making this comparison, both quantities must be expressed in terms of the same units.

Example:

Express the ratio of 1 hour to 1 day.

Solution:

A day contains 24 hours. The ratio is $\dfrac{1}{24}$, which can also be written 1 : 24.

Example:

Find the ratio of the shaded portion to the unshaded portion.

Solution:

There are 5 squares shaded out of 9. The ratio of the shaded portion to unshaded portion is $\dfrac{5}{4}$.

A proportion is a statement of equality between two ratios. The denominator of the first fraction and the numerator of the second are called the means of the proportion. The numerator of the first fraction and the denominator of the second are called the extremes. In solving a proportion, we use the theorem that states the product of the means is equal to the product of the extremes. We refer to this as *cross multiplying*.

Example:

Solve for *x:* $\dfrac{x+3}{5} = \dfrac{8-x}{6}$

Solution:

Cross multiply. $6x + 18 = 40 - 5x$

$$11x = 22$$
$$x = 2$$

Example:

Solve for *x:* 4 : *x* = 9 : 18

Solution:

Rewrite in fraction form. $\dfrac{4}{x} = \dfrac{9}{18}$
Cross multiply. $9x = 72$

$$x = 8$$

If you observe that the second fraction is equal to $\dfrac{1}{2}$, then the first must also be equal to $\dfrac{1}{2}$. Therefore, the missing denominator must be 8. Observation often saves valuable time.

Exercise 1

Work out each problem. Circle the letter that appears before your answer.

1. Find the ratio of 1 ft. 4 in. to 1 yd.

   (A) $1 : 3$
   (B) $2 : 9$
   (C) $4 : 9$
   (D) $3 : 5$
   (E) $5 : 12$

2. A team won 25 games in a 40 game season. Find the ratio of games won to games lost.

   (A) $\dfrac{5}{8}$
   (B) $\dfrac{3}{8}$
   (C) $\dfrac{3}{5}$
   (D) $\dfrac{5}{3}$
   (E) $\dfrac{3}{2}$

3. In the proportion $a : b = c : d$, solve for d in terms of a, b and c.

   (A) $\dfrac{ac}{b}$
   (B) $\dfrac{bc}{a}$
   (C) $\dfrac{ab}{c}$
   (D) $\dfrac{a}{bc}$
   (E) $\dfrac{bc}{d}$

4. Solve for x: $\dfrac{x+1}{8} = \dfrac{28}{32}$

   (A) $6\dfrac{1}{2}$
   (B) 5
   (C) 4
   (D) 7
   (E) 6

5. Solve for y: $\dfrac{2y}{9} = \dfrac{y-1}{3}$

   (A) 3
   (B) $\dfrac{1}{3}$
   (C) $\dfrac{9}{15}$
   (D) $\dfrac{9}{4}$
   (E) $\dfrac{4}{9}$

2. DIRECT VARIATION

Two quantities are said to vary directly if they change in the same direction. As the first increases, the second does also. As the first decreases, the second does also.

For example, the distance you travel at a constant rate varies directly as the time spent traveling. The number of pounds of apples you buy varies directly as the amount of money you spend. The number of pounds of butter you use in a cookie recipe varies directly as the number of cups of sugar you use.

Whenever two quantities vary directly, a problem can be solved using a proportion. We must be very careful to compare quantities in the same order and in terms of the same units in both fractions. If we compare miles with hours in the first fraction, we must compare miles with hours in the second fraction.

You must always be sure that as one quantity increases or decreases, the other changes in the same direction before you try to solve using a proportion.

Example:

If 4 bottles of milk cost $2, how many bottles of milk can you buy for $8?

Solution:

The more milk you buy, the more it will cost. This is *direct*. We are comparing the number of bottles with cost.

$$\frac{4}{2} = \frac{x}{8}$$

If we cross multiply, we get $2x = 32$ or $x = 16$.

A shortcut in the above example would be to observe what change takes place in the denominator and apply the same change to the numerator. The denominator of the left fraction was multiplied by 4 to give the denominator of the right fraction. Therefore we multiply the numerator by 4 as well to maintain the equality. This method often means a proportion can be solved at sight with no written computation at all, saving valuable time.

Example:

If b boys can deliver n newspapers in one hour, how many newspapers can c boys deliver in the same time?

Solution:

The more boys, the more papers will be delivered. This is *direct*. We are comparing the number of boys with the number of newspapers.

$$\frac{b}{n} = \frac{c}{x}$$ Cross multiply and solve for x.

$$bx = cn$$

$$x = \frac{cn}{b}$$

Exercise 2

Work out each problem. Circle the letter that appears before your answer.

1. Find the cost, in cents, of 8 books if 3 books of the same kind cost D dollars.

   (A) $\dfrac{8D}{3}$

   (B) $\dfrac{3}{800D}$

   (C) $\dfrac{3}{8D}$

   (D) $\dfrac{800D}{3}$

   (E) $\dfrac{108D}{3}$

2. On a map $\dfrac{1}{2}$ inch = 10 miles. How many miles apart are two towns that are $2\dfrac{1}{4}$ inches apart on the map?

   (A) $11\dfrac{1}{4}$

   (B) 45

   (C) $22\dfrac{1}{2}$

   (D) $40\dfrac{1}{2}$

   (E) 42

3. The toll on the Intercoastal Thruway is 8¢ for every 5 miles traveled. What is the toll for a trip of 115 miles on this road?

   (A) $9.20
   (B) $1.70
   (C) $1.84
   (D) $1.64
   (E) $1.76

4. Mark's car uses 20 gallons of gas to drive 425 miles. At this rate, approximately how many gallons of gas will he need for a trip of 1000 miles?

   (A) 44
   (B) 45
   (C) 46
   (D) 47
   (E) 49

5. If r planes can carry p passengers, how many planes are needed to carry m passengers?

   (A) $\dfrac{m}{p}$

   (B) $\dfrac{rp}{m}$

   (C) $\dfrac{p}{rm}$

   (D) $\dfrac{pm}{r}$

   (E) $\dfrac{m}{rp}$

3. INVERSE VARIATION

Two quantities are said to vary inversely if they change in opposite directions. As the first increases, the second decreases. As the first decreases, the second increases.

Whenever two quantities vary inversely, their product remains constant. Instead of dividing one quantity by the other and setting their quotients equal as we did in direct variation, we multiply one quantity by the other and set the products equal.

There are several situations that are good examples of inverse variation.

A) The number of teeth in a meshed gear varies inversely as the number of revolutions it makes per minute. The more teeth a gear has, the fewer revolutions it will make per minute. The less teeth it has, the more revolutions it will make per minute. The product of the number of teeth and the revolutions per minute remains constant.

B) The distance a weight is placed from the fulcrum of a balanced lever varies inversely as its weight. The heavier the object, the shorter must be its distance from the fulcrum. The lighter the object, the greater must be the distance. The product of the weight of the object and its distance from the fulcrum remains constant.

C) When two pulleys are connected by a belt, the diameter of a pulley varies inversely as the number of revolutions per minute. The larger the diameter, the smaller the number of revolutions per minute. The smaller the diameter, the greater the number of revolutions per minute. The product of the diameter of a pulley and the number of revolutions per minute remains constant.

D) The number of people hired to work on a job varies inversely as the time needed to complete the job. The more people working, the less time it will take. The fewer people working, the longer it will take. The product of the number of people and the time worked remains constant.

E) How long food, or any commodity, lasts varies inversely as the number of people who consume it. The more people, the less time it will last. The fewer people, the longer it will last. The product of the number of people and the time it will last remains constant.

Example:

> If 3 men can paint a house in 2 days, how long will it take 2 men to do the same job?

Solution:

> The fewer men, the more days. This is *inverse*.
>
> $3 \cdot 2 = 2 \cdot x$
>
> $6 = 2x$
>
> $x = 3$ days

Exercise 3

Work out each problem. Circle the letter that appears before your answer.

1. A field can be plowed by 8 machines in 6 hours. If 3 machines are broken and cannot be used, how many hours will it take to plow the field?

   (A) 12
   (B) $9\frac{3}{5}$
   (C) $3\frac{3}{4}$
   (D) 4
   (E) 16

2. Camp Starlight has enough milk to feed 90 children for 4 days. If 10 of the children do not drink milk, how many days will the supply last?

   (A) 5
   (B) 6
   (C) $4\frac{1}{2}$
   (D) $4\frac{1}{8}$
   (E) $5\frac{1}{3}$

3. A pulley revolving at 200 revolutions per minute has a diameter of 15 inches. It is belted to a second pulley which revolves at 150 revolutions per minute. Find the diameter, in inches, of the second pulley.

   (A) 11.2
   (B) 20
   (C) 18
   (D) 16.4
   (E) 2

4. Two boys weighing 60 pounds and 80 pounds balance a seesaw. How many feet from the fulcrum must the heavier boy sit if the lighter boy is 8 feet from the fulcrum?

   (A) 10
   (B) $10\frac{2}{3}$
   (C) 9
   (D) $7\frac{1}{2}$
   (E) 6

5. A gear with 20 teeth revolving at 200 revolutions per minute is meshed with a second gear turning at 250 revolutions per minute. How many teeth does this gear have?

   (A) 16
   (B) 25
   (C) 15
   (D) 10
   (E) 24

In solving variation problems, you must decide whether the two quantities involved change in the same direction, in which case it is direct variation and should be solved by means of proportions. If the quantities change in opposite directions, it is inverse variation, solved by means of constant products. In the following exercises, decide carefully whether each is an example of direct or inverse variation.

Exercise 4

Work out each problem. Circle the letter that appears before your answer.

1. A farmer has enough chicken feed to last 30 chickens for 4 days. If 10 more chickens are added, how many days will the feed last?

   (A) 3

   (B) $1\frac{1}{3}$

   (C) 12

   (D) $2\frac{2}{3}$

   (E) $5\frac{1}{3}$

2. At c cents per can, what is the cost of p cases of soda if there are 12 cans in a case?

   (A) $12cp$

   (B) $\dfrac{cp}{12}$

   (C) $\dfrac{12}{cp}$

   (D) $\dfrac{12p}{c}$

   (E) $\dfrac{12c}{p}$

3. If m boys can put up a fence in d days, how many days will it take to put up the fence if two of the boys cannot participate?

   (A) $\dfrac{d}{-2}$

   (B) $\dfrac{d(m-2)}{m}$

   (C) $\dfrac{md}{m-2}$

   (D) $\dfrac{m-2}{md}$

   (E) $\dfrac{m(m-2)}{d}$

4. A recipe calls for $\frac{3}{4}$ lb. of butter and 18 oz. of sugar. If only 10 oz. of butter are available, how many ounces of sugar should be used?

   (A) $13\frac{1}{2}$

   (B) 23

   (C) 24

   (D) 14

   (E) 15

5. If 3 kilometers are equal to 1.8 miles, how many kilometers are equal to 100 miles?

   (A) 60

   (B) $166\frac{2}{3}$

   (C) 540

   (D) $150\frac{1}{2}$

   (E) 160.4

RETEST

Work out each problem. Circle the letter that appears before your answer.

1. Solve for x: $\dfrac{3x}{8} = \dfrac{x+7}{12}$

   (A) $\dfrac{7}{28}$
   (B) 2
   (C) 4
   (D) $2\dfrac{3}{4}$
   (E) 1

2. Solve for x if $a = 5$, $b = 8$, and $c = 3$: $\dfrac{a-3}{x} = \dfrac{b+2}{5c}$

   (A) 5
   (B) 20
   (C) 2
   (D) 3
   (E) 6

3. A map is drawn to a scale of $\dfrac{1}{2}$ inch = 20 miles. How many miles apart are two cities that are $3\dfrac{1}{4}$ inches apart on the map?

   (A) 70
   (B) 130
   (C) 65
   (D) $32\dfrac{1}{2}$
   (E) 35

4. Mr. Weiss earned $12,000 during the first 5 months of the year. If his salary continues at the same rate, what will his annual income be that year?

   (A) $60,000
   (B) $28,000
   (C) $27,000
   (D) $30,000
   (E) $28,800

5. How many pencils can be bought for D dollars if n pencils cost c cents?

   (A) $\dfrac{nD}{c}$
   (B) $\dfrac{nD}{100c}$
   (C) $\dfrac{100D}{nc}$
   (D) $\dfrac{100nD}{c}$
   (E) $\dfrac{nc}{100D}$

6. Ten boys agree to paint the gym in 5 days. If five more boys join in before the work begins, how many days should the painting take?

   (A) $3\dfrac{1}{3}$
   (B) $3\dfrac{1}{2}$
   (C) 10
   (D) $2\dfrac{1}{2}$
   (E) $2\dfrac{3}{4}$

7. A weight of 120 pounds is placed five feet from the fulcrum of a lever. How far from the fulcrum should a 100 pound weight be placed in order to balance the lever?

   (A) 6 ft.
   (B) $4\dfrac{1}{6}$ ft.
   (C) $5\dfrac{1}{2}$ ft.
   (D) $6\dfrac{1}{2}$ ft.
   (E) $6\dfrac{2}{3}$ ft.

8. A photograph negative measures $1\dfrac{7}{8}$ inches by $2\dfrac{1}{2}$ inches. The printed picture is to have its longer dimension be 4 inches. How long should the shorter dimension be?

   (A) $2\dfrac{3}{8}''$
   (B) $2\dfrac{1}{2}''$
   (C) $3''$
   (D) $3\dfrac{1}{8}''$
   (E) $3\dfrac{3}{8}''$

9. A gear with 60 teeth is meshed to a gear with 40 teeth. If the larger gear revolves at 20 revolutions per minute, how many revolutions does the smaller gear make in a minute?

   (A) $13\frac{1}{3}$
   (B) 3
   (C) 300
   (D) 120
   (E) 30

10. How many gallons of paint must be purchased to paint a room containing 820 square feet of wall space, if one gallon covers 150 square feet? (Any fraction must be rounded *up*.)

   (A) 4
   (B) 5
   (C) 6
   (D) 7
   (E) 8

SOLUTIONS TO PRACTICE EXERCISES

Diagnostic Test

1. (B) $2x(4) = 3(x + 5)$
    $$8x = 3x + 15$$
    $$5x = 15$$
    $$x = 3$$

2. (E) $\dfrac{4}{x} = \dfrac{10}{20}$ Cross multiply.
    $$80 = 10x$$
    $$x = 8$$

3. (C) We compare inches to miles.

    $\dfrac{2}{25} = \dfrac{5\frac{2}{5}}{x}$ Cross multiply.

    $$2x = 135$$
    $$x = 67\frac{1}{2}$$

4. (A) We compare apples to cents.

    $$\frac{x}{c} = \frac{n}{d}$$
    $$dx = nc$$ Cross multiply.
    $$x = \frac{nc}{d}$$

5. (B) We compare miles to months.
    $$\frac{5}{7000} = \frac{12}{x}$$
    $$5x = 84{,}000$$
    $$x = 16{,}800$$

6. (E) Number of teeth times speed remains constant.
    $$20 \cdot 30 = x \cdot 25$$
    $$600 = 25x$$
    $$x = 24$$

7. (B) Weight times distance from the fulcrum remains constant.
    $$90 \cdot 3 = 50 \cdot x$$
    $$270 = 50x$$
    $$x = 5\frac{2}{5} \text{ ft.}$$

8. (C) The more dogs, the fewer days. This is inverse variation.
    $$2 \cdot 3 = 3 \cdot x$$
    $$6 = 3x$$
    $$x = 2 \text{ weeks} = 14 \text{ days}$$

9. (D) Number of machines times hours needed remains constant.
    $$m \cdot h = (m - 2) \cdot x$$
    $$x = \frac{mh}{m - 2}$$

10. (C) The more men, the fewer days. This is inverse variation.
    $$20 \cdot 6 = 24 \cdot x$$
    $$120 = 24x$$
    $$x = 5$$

    The rations will last 1 day less.

Exercise 1

1. **(C)** 1 ft. 4 in. = 16 in.

   1 yd. = 36 in.

   $$\frac{16}{36} = \frac{4}{9}$$

2. **(D)** The team won 25 games and lost 15.

   $$\frac{25}{15} = \frac{5}{3}$$

3. **(B)** $\dfrac{a}{b} = \dfrac{c}{d}$ Cross multiply. Divide by a.

   $$ad = bc$$
   $$d = \frac{bc}{a}$$

4. **(E)** $32(x + 1) = 28(8)$

   $$32x + 32 = 224$$
   $$32x = 192$$
   $$x = 6$$

5. **(A)** $9(y - 1) = 2y(3)$

   $$9y - 9 = 6y$$
   $$3y = 9$$
   $$y = 3$$

Exercise 2

1. **(D)** We compare books with cents. D dollars is equivalent to $100D$ cents.

   $$\frac{3}{100D} = \frac{8}{x}$$
   $$3x = 800D$$
   $$x = \frac{800D}{3}$$

2. **(B)** We compare inches to miles.

   $$\frac{\frac{1}{2}}{10} = \frac{2\frac{1}{4}}{x}$$
   $$\frac{1}{2}x = 22\frac{1}{2} \quad \text{Cross multiply. Multiply by 2.}$$
   $$x = 45$$

3. **(C)** We compare cents to miles.

   $$\frac{8}{5} = \frac{x}{115}$$
   $$5x = 920 \quad \text{Cross multiply.}$$
   $$x = \$1.84$$

4. **(D)** We compare gallons to miles.

   $$\frac{20}{425} = \frac{x}{1000} \quad \text{Cross multiply. To avoid}$$
   $$425x = 20{,}000 \quad \text{large numbers, divide by 25.}$$

   $$17x = 800$$
   $$x = 47\frac{1}{17}$$

5. **(A)** We compare planes to passengers.

   $$\frac{r}{p} = \frac{x}{m} \quad \text{Cross multiply. Divide by } p.$$
   $$px = rm$$
   $$x = \frac{rm}{p}$$

Exercise 3

1. (B) Number of machines times hours needed remains constant.

$$8 \cdot 6 = 5 \cdot x$$
$$48 = 5x$$
$$x = 9\frac{3}{5}$$

2. (C) Number of children times days remains constant.

$$90 \cdot 4 = 80 \cdot x$$
$$80x = 360$$
$$x = 4\frac{1}{2}$$

3. (B) Diameter times speed remains constant.

$$15 \cdot 200 = x \cdot 150$$
$$3000 = 150x$$
$$x = 20$$

4. (E) Weight times distance from fulcrum remains constant.

$$80 \cdot x = 60 \cdot 8$$
$$80x = 480$$
$$x = 6$$

5. (A) Number of teeth times speed remains constant.

$$20 \cdot 200 = x \cdot 250$$
$$250x = 4000$$
$$x = 16$$

Exercise 4

1. (A) The more chickens, the fewer days. This is *inverse*.

$$30 \cdot 4 = 40 \cdot x$$
$$40x = 120$$
$$x = 3$$

2. (A) The more cases, the more cents. This is *direct*. We compare cents with cans. In p cases there will be $12p$ cans.

$$\frac{c}{1} = \frac{x}{12p}$$
$$x = 12cp$$

3. (C) The fewer boys, the more days. This is *inverse*.

$$m \cdot d = (m-2) \cdot x$$
$$\frac{md}{m-2} = x$$

4. (E) The less butter, the less sugar. This is *direct*. Change $\frac{3}{4}$ lb. to 12 oz.

$$\frac{12}{18} = \frac{10}{x}$$
$$12x = 180$$
$$x = 15$$

5. (B) The more kilometers, the more miles. This is *direct*.

$$\frac{3}{1.8} = \frac{x}{100}$$
$$1.8x = 300$$
$$18x = 3000$$
$$x = 166\frac{2}{3}$$

Retest

1. (B) $3x(12) = 8(x + 7)$

$$36x = 8x + 56$$
$$28x = 56$$
$$x = 2$$

2. (D) $\dfrac{2}{x} = \dfrac{10}{15}$ Cross multiply.

$$30 = 10x$$
$$x = 3$$

3. (B) We compare inches to miles.

$$\dfrac{\frac{1}{2}}{20} = \dfrac{3\frac{1}{4}}{x}$$ Cross multiply. Multiply by 2.

$$\frac{1}{2}x = 65$$
$$x = 130$$

4. (E) We compare dollars to months.

$$\dfrac{12,000}{5} = \dfrac{x}{12}$$ Cross multiply.

$$144,000 = 5x$$
$$x = \$28,800$$

5. (D) We compare pencils to dollars. The cost of n pencils is $\dfrac{c}{100}$ dollars.

$$\dfrac{x}{D} = \dfrac{n}{\frac{c}{100}}$$ Cross multiply. Multiply by $\dfrac{100}{c}$.

$$\dfrac{cx}{100} = nD$$
$$x = \dfrac{100nD}{c}$$

6. (A) The more boys, the fewer days. This is *inverse*.

$$10 \cdot 5 = 15 \cdot x$$
$$15x = 50$$
$$x = 3\frac{1}{3}$$

7. (A) Weight times distance from the fulcrum remains constant.

$$120 \cdot 5 = 100 \cdot x$$
$$600 = 100x$$
$$x = 6 \text{ ft.}$$

8. (C)

$$\dfrac{2\frac{1}{2}}{4} = \dfrac{1\frac{7}{8}}{x}$$ Cross multiply. Multiply by 2.

$$\frac{5}{2}x = \frac{15}{2}$$
$$5x = 15$$
$$x = 3''$$

9. (E) Number of teeth times speed remains constant.

$$60 \cdot 20 = 40 \cdot x$$
$$1200 = 40x$$
$$x = 30$$

10. (C) We compare gallons to square feet.

$$\dfrac{x}{820} = \dfrac{1}{150}$$ Cross multiply.

$$150x = 820$$

$x = 5.47$, which means 6 gallons must be purchased

Percent

5

DIAGNOSTIC TEST

Directions: Work out each problem. Circle the letter that appears before your answer.

Answers are at the end of the chapter.

1. Write as a fraction: 4.5%

   (A) $\dfrac{9}{2}$

   (B) $\dfrac{9}{20}$

   (C) $\dfrac{9}{200}$

   (D) $\dfrac{9}{2000}$

   (E) $\dfrac{4.5}{10}$

2. Write $\dfrac{2}{5}$ % as a decimal.

   (A) .40
   (B) .04
   (C) 40.0
   (D) .004
   (E) 4.00

3. What is $62\dfrac{1}{2}$ % of 80?

   (A) 5000
   (B) 500
   (C) 50
   (D) 5
   (E) .5

4. Find 6% of b.

   (A) .6b
   (B) .06b

   (C) $\dfrac{b}{6}$

   (D) $\dfrac{b}{.06}$

   (E) $\dfrac{100b}{6}$

5. 80 is 40% of what number?

   (A) 3200
   (B) 320
   (C) 32
   (D) 200
   (E) 20

6. c is $83\dfrac{1}{3}$ % of what number?

   (A) $\dfrac{5c}{6}$

   (B) $\dfrac{6c}{5}$

   (C) $\dfrac{7c}{8}$

   (D) $\dfrac{8c}{7}$

   (E) $\dfrac{2c}{3}$

7. How many sixteenths are there in $87\dfrac{1}{2}$ %?

   (A) 7
   (B) 8
   (C) 10
   (D) 12
   (E) 14

8. What percent of 40 is 16?

   (A) $2\dfrac{1}{2}$
   (B) 25
   (C) 30
   (D) 40
   (E) 45

9. Find 112% of 80.
   (A) 92
   (B) 89.6
   (C) 88
   (D) 70.5
   (E) 91

10. What percent of 60 is 72?
    (A) 105
    (B) 125
    (C) 120
    (D) $83\frac{1}{3}$
    (E) 110

1. FRACTIONAL AND DECIMAL EQUIVALENTS OF PERCENTS

Percent means "out of 100." If you understand this concept, it then becomes very easy to change a percent to an equivalent decimal or fraction.

Example:

5% means 5 out of 100 or $\dfrac{5}{100}$, which is equal to .05

3.4% means 3.4 out of 100 or $\dfrac{3.4}{100}$, which is equivalent to $\dfrac{34}{1000}$ or .034

c% means c out of 100 or $\dfrac{c}{100}$, which is equivalent to $\dfrac{1}{100} \cdot c$ or .01c

$\dfrac{1}{4}$% means $\dfrac{1}{4}$ out of 100 or $\dfrac{\frac{1}{4}}{100}$, which is equivalent to $\dfrac{1}{100} \cdot .25$ or .0025

To change a percent to a decimal, therefore, we must move the decimal point two places to the *left,* as we are dividing by 100.

Example:

62% = .62
.4% = .004
3.2% = .032

To change a decimal to a percent, we must reverse the above steps. We multiply by 100, which has the effect of moving the decimal point two places to the *right*, and insert the percent sign.

Example:

.27 = 27%
.012 = 1.2%
.003 = .3%

To change a percent to a fraction, we remove the percent sign and divide by 100. This has the effect of putting the percent over 100 and then simplifying the resulting fraction.

Example:

$25\% = \dfrac{25}{100} = \dfrac{1}{4}$

$70\% = \dfrac{70}{100} = \dfrac{7}{10}$

$.5\% = \dfrac{.5}{100} = \dfrac{5}{1000} = \dfrac{1}{200}$

To change a fraction to a percent, we must reverse the above steps. We multiply by 100 and insert the percent sign.

Example:

$\dfrac{4}{5} = \dfrac{4}{\cancel{5}} \cdot \cancel{100}^{\,20} \% = 80\%$

$\dfrac{3}{8} = \dfrac{3}{\cancel{8}_{2}} \cdot \cancel{100}^{\,25} \% = \dfrac{75}{2}\% = 37\dfrac{1}{2}\%$

Some fractions do not convert easily, as the denominator does not divide into 100. Such fractions must be changed to decimals first by dividing the numerator by the denominator. Then convert the decimal to a percent as explained on the previous page. Divide to two places only, unless it clearly comes out even in one or two additional places.

Example:

$$\frac{8}{17} = 17\overline{)8.00}^{.47} = 47\frac{1}{17}\%$$

$$\begin{array}{r} 6\ 8 \\ \hline 1\ 20 \\ 1\ 19 \\ \hline 1 \end{array}$$

$$\frac{4}{125} = 125\overline{)4.000}^{.032} = 3.2\%$$

$$\begin{array}{r} 3\ 75 \\ \hline 250 \\ 250 \\ \hline \end{array}$$

Certain fractional and decimal equivalents of common percents occur frequently enough so that they should be memorized. Learning the values in the following table will make your work with percent problems much easier.

PERCENT	DECIMAL	FRACTION
50%	.5	$\frac{1}{2}$
25%	.25	$\frac{1}{4}$
75%	.75	$\frac{3}{4}$
10%	.1	$\frac{1}{10}$
30%	.3	$\frac{3}{10}$
70%	.7	$\frac{7}{10}$
90%	.9	$\frac{9}{10}$
$33\frac{1}{3}\%$	$.3\overline{3}$	$\frac{1}{3}$
$66\frac{2}{3}\%$	$.6\overline{6}$	$\frac{2}{3}$
$16\frac{2}{3}\%$	$.1\overline{6}$	$\frac{1}{6}$
$83\frac{1}{3}\%$	$.8\overline{3}$	$\frac{5}{6}$
20%	.2	$\frac{1}{5}$
40%	.4	$\frac{2}{5}$
60%	.6	$\frac{3}{5}$
80%	.8	$\frac{4}{5}$
$12\frac{1}{2}\%$	.125	$\frac{1}{8}$
$37\frac{1}{2}\%$	.375	$\frac{3}{8}$
$62\frac{1}{2}\%$	.625	$\frac{5}{8}$
$87\frac{1}{2}\%$	.875	$\frac{7}{8}$

Exercise 1

Work out each problem. Circle the letter that appears before your answer.

1. $3\frac{1}{2}\%$ may be written as a decimal as

   (A) 3.5
   (B) .35
   (C) .035
   (D) .0035
   (E) 3.05

2. Write as a fraction in simplest form: 85%.

   (A) $\frac{13}{20}$
   (B) $\frac{17}{20}$
   (C) $\frac{17}{10}$
   (D) $\frac{19}{20}$
   (E) $\frac{17}{2}$

3. Write 4.6 as a percent.

   (A) 4.6%
   (B) .46%
   (C) .046%
   (D) 46%
   (E) 460%

4. Write $\frac{5}{12}$ as an equivalent percent.

   (A) 41%
   (B) 41.6%
   (C) $41\frac{2}{3}\%$
   (D) 4.1%
   (E) $.41\frac{2}{3}\%$

5. Write $\frac{1}{2}\%$ as a decimal.

   (A) .5
   (B) .005
   (C) 5.0
   (D) 50.0
   (E) .05

2. FINDING A PERCENT OF A NUMBER

Most percentage problems can be solved by using the proportion

$$\frac{\%}{100} = \frac{\text{part}}{\text{whole}}.$$

Although this method will work, it often yields unnecessarily large numbers that make for difficult computation. As we look at different types of percent problems, we will compare methods of solution. In finding a percent of a number, it is usually easier to change the percent to an equivalent decimal or fraction and multiply by the given number.

Example:

Find 32% of 84.

Proportion Method

$$\frac{32}{100} = \frac{x}{84}$$
$$100x = 2688$$
$$x = 26.88$$

Decimal Method
Change 32% to .32 and multiply.

$$\begin{array}{r} 84 \\ \times\, .32 \\ \hline 168 \\ 252 \\ \hline 26.88 \end{array}$$

Example:

Find $12\frac{1}{2}\%$ of 112.

Proportion Method

$$\frac{12\frac{1}{2}}{100} = \frac{x}{112}$$
$$100x = 1400$$
$$x = 14$$

Decimal Method

$$\begin{array}{r} 112 \\ \times\, .125 \\ \hline 560 \\ 2\,24 \\ 11\,2 \\ \hline 14.000 \end{array}$$

Fraction Method

Change $12\frac{1}{2}\%$ to $\frac{1}{8}$

$$\frac{1}{\cancel{8}} \cdot \cancel{112}^{14} = 14$$

Which method do you think is the easiest? When the fractional equivalent of the required percent is among those given in the previous chart, the fraction method is by far the least time-consuming. It really pays to memorize those fractional equivalents.

Exercise 2

Work out each problem. Circle the letter that appears before your answer.

1. What is 40% of 40?

   (A) .16
   (B) 1.6
   (C) 16
   (D) 160
   (E) 1600

2. What is 42% of 67?

   (A) 2814
   (B) 281.4
   (C) 2.814
   (D) .2814
   (E) 28.14

3. Find $16\frac{2}{3}$% of 120.

   (A) 20
   (B) 2
   (C) 200
   (D) 16
   (E) 32

4. What is $\frac{1}{5}$% of 40?

   (A) 8
   (B) .8
   (C) .08
   (D) .008
   (E) .0008

5. Find r% of s.

   (A) $\dfrac{100s}{r}$

   (B) $\dfrac{rs}{100}$

   (C) $\dfrac{100r}{s}$

   (D) $\dfrac{r}{100s}$

   (E) $\dfrac{s}{100r}$

3. FINDING A NUMBER WHEN A PERCENT OF IT IS GIVEN

This type of problem may be solved using the proportion method, although this may again result in the unnecessary use of time. It is often easier to translate the words of such a problem into an algebraic statement, using decimal or fractional equivalents for the percents involved. Then it will become evident that we divide the given number by the given percent to solve.

Example:

7 is 5% of what number?

Proportion Method

$$\frac{5}{100} = \frac{7}{x}$$
$$5x = 700$$
$$x = 140$$

Equation Method

$$7 = .05x$$
$$700 = 5x$$
$$140 = x$$

Example:

40 is $66\frac{2}{3}$% of what number?

Proportion Method

$$\frac{66\frac{2}{3}}{100} = \frac{40}{x}$$
$$66\frac{2}{3}x = 4000$$
$$\frac{200}{3}x = 4000$$
$$200x = 12000$$
$$2x = 120$$
$$x = 60$$

Equation Method

$$40 = \frac{2}{3}x$$
$$120 = 2x$$
$$60 = x$$

Just think of the amount of time you will save and the extra problems you will get to do if you know that $66\frac{2}{3}$% is $\frac{2}{3}$ and use the equation method. Are you convinced that the common fraction equivalents in the previously given chart should be memorized?

Exercise 3

Work out each problem. Circle the letter that appears before your answer.

1. 72 is 12% of what number?
   - (A) 6
   - (B) 60
   - (C) 600
   - (D) 86.4
   - (E) 8.64

2. 80 is $12\frac{1}{2}$ % of what number?
   - (A) 10
   - (B) 100
   - (C) 64
   - (D) 640
   - (E) 6400

3. $37\frac{1}{2}$ % of what number is 27?
   - (A) 72
   - (B) $10\frac{1}{8}$
   - (C) 90
   - (D) 101.25
   - (E) 216

4. m is p% of what number?
   - (A) $\dfrac{mp}{100}$
   - (B) $\dfrac{100p}{m}$
   - (C) $\dfrac{m}{100p}$
   - (D) $\dfrac{p}{100m}$
   - (E) $\dfrac{100m}{p}$

5. 50% of what number is r?
   - (A) $\dfrac{1}{2} r$
   - (B) $5r$
   - (C) $10r$
   - (D) $2r$
   - (E) $100r$

4. TO FIND WHAT PERCENT ONE NUMBER IS OF ANOTHER

This type of problem may also be solved using the proportion method. However, this may again result in the use of an unnecessary amount of time. It is often easier to put the part over the whole, simplify the resulting fraction, and multiply by 100.

Example:

30 is what percent of 1500?

Proportion Method

$$\frac{x}{100} = \frac{30}{1500}$$
$$1500x = 3000$$
$$x = 2\%$$

Fraction Method

$$\frac{30}{1500} = \frac{3}{150} = \frac{1}{50} \cdot 100 = 2\%$$

Example:

12 is what percent of 72?

Proportion Method

$$\frac{x}{100} = \frac{12}{72}$$
$$72x = 1200$$

Fraction Method

$$\frac{12}{72} = \frac{1}{6} = 16\frac{2}{3}\%$$

Time consuming long division is needed to find $x = 16\frac{2}{3}\%$. If you have memorized the fractional equivalents of common percents, this method requires only a few seconds.

Example:

What percent of 72 is 16?

Proportion Method

$$\frac{x}{100} = \frac{16}{72}$$
$$72x = 1600$$
$$x = 22\frac{2}{9}\%$$

Fraction Method

$$\frac{16}{72} = \frac{2}{9} \cdot 100 = \frac{200}{9} = 22\frac{2}{9}\%$$

Exercise 4

Work out each problem. Circle the letter that appears before your answer.

1. 4 is what percent of 80?

   (A) 20
   (B) 2
   (C) 5
   (D) .5
   (E) 40

2. $\frac{1}{2}$ of 6 is what percent of $\frac{1}{4}$ of 60?

   (A) 5
   (B) 20
   (C) 10
   (D) 25
   (E) 15

3. What percent of 96 is 12?

   (A) $16\frac{2}{3}$

   (B) $8\frac{1}{3}$

   (C) $37\frac{1}{2}$

   (D) 8

   (E) $12\frac{1}{2}$

4. What percent of 48 is 48?

   (A) 1
   (B) 10
   (C) 100
   (D) 48
   (E) 0

5. What percent of y is x?

   (A) $\dfrac{x}{y}$

   (B) $\dfrac{x}{100y}$

   (C) $\dfrac{xy}{100}$

   (D) $\dfrac{100x}{y}$

   (E) $\dfrac{100y}{x}$

5. PERCENTS GREATER THAN 100

When the percentage involved in a problem is greater than 100, the same methods apply. Remember that 100% = 1; 200% = 2; 300% = 3 and so forth. Therefore 150% will be equal to 100% + 50% or $1\frac{1}{2}$. Let us look at one example of each previously discussed problem, using percents greater than 100.

Example:

Find 175% of 60

Proportion Method	*Decimal Method*	*Fraction Method*
	60	
	$\times\ 1.75$	
$\frac{175}{100}=\frac{x}{60}$	300	$1\frac{3}{4}\cdot 60$
$100x=10500$	4200	$\frac{7}{\cancel{4}}\cdot\overset{15}{\cancel{60}}=105$
$x=105$	$\underline{6000}$	
	105.00	

Example:

80 is 125% of what number?

Proportion Method	*Decimal Method*	*Fraction Method*
		$80=1\frac{1}{4}x$
$\frac{125}{100}=\frac{80}{x}$	$80=1.25x$	$80=\frac{5}{4}x$
$125x=8000$	$8000=125x$	$320=5x$
$x=64$	$x=64$	$x=64$

Example:

40 is what percent of 30?

Proportion Method	*Fraction Method*
$\frac{x}{100}=\frac{40}{30}$	
$30x=4000$	$\frac{40}{30}=\frac{4}{3}=1\frac{1}{3}=133\frac{1}{3}\%$
$x=133\frac{1}{3}\%$	

Exercise 5

Work out each problem. Circle the letter that appears before your answer.

1. 36 is 150% of what number?

   (A) 24
   (B) 54
   (C) 26
   (D) 12
   (E) 48

2. What is 300% of 6?

   (A) 2
   (B) 3
   (C) 12
   (D) 18
   (E) 24

3. What percent of 90 is 120?

   (A) 75
   (B) $133\frac{1}{3}$
   (C) 125
   (D) 120
   (E) $1\frac{1}{3}$

4. 500 is 200% of what number?

   (A) 250
   (B) 1000
   (C) 100
   (D) 750
   (E) 300

5. To multiply a number by $137\frac{1}{2}$%, the number should be multiplied by

   (A) 137.5
   (B) 13750
   (C) 1.375
   (D) 13.75
   (E) .1375

RETEST

Work out each problem. Circle the letter that appears before your answer.

1. Write as a fraction in lowest terms: .25%.

   (A) $\frac{1}{4}$

   (B) $\frac{1}{40}$

   (C) $\frac{1}{400}$

   (D) $\frac{1}{4000}$

   (E) $\frac{1}{25}$

2. Write $\frac{3}{4}$% as a decimal.

   (A) .75
   (B) 75.0
   (C) .075
   (D) .0075
   (E) 7.5

3. Find 12% of 80.

   (A) 10
   (B) .96
   (C) .096
   (D) 960
   (E) 9.6

4. 18 is 20% of what number?

   (A) 3.6
   (B) 90
   (C) 72
   (D) 21.6
   (E) 108

5. What is b% of 6?

   (A) $\frac{3b}{50}$

   (B) $\frac{3}{50b}$

   (C) $\frac{50b}{3}$

   (D) $\frac{50}{3b}$

   (E) $\frac{b}{150}$

6. m is $62\frac{1}{2}$% of what number?

   (A) $\frac{5m}{8}$

   (B) $\frac{8m}{5}$

   (C) $8m$

   (D) $\frac{5}{8m}$

   (E) $\frac{8}{5m}$

7. What percent of 12 is 2?

   (A) 600

   (B) $12\frac{1}{2}$

   (C) $16\frac{2}{3}$

   (D) $6\frac{2}{3}$

   (E) 6

8. What is 140% of 70?

   (A) 9800
   (B) 980
   (C) .98
   (D) 9.8
   (E) 98

9. How many fifths are there in 280%?

   (A) 28
   (B) 1.4
   (C) 14
   (D) 56
   (E) 2.8

10. What percent of 12 is 16?

   (A) $133\frac{1}{3}$

   (B) 125
   (C) 75
   (D) 80

   (E) $1\frac{1}{4}$

SOLUTIONS TO PRACTICE EXERCISES

Diagnostic Test

1. (C) $4.5\% = \dfrac{4.5}{100} = \dfrac{45}{1000} = \dfrac{9}{200}$

2. (D) $\dfrac{2}{5}\% = .4\% = .004$

3. (C) $62\dfrac{1}{2}\% = \dfrac{5}{8} \quad \dfrac{5}{\cancel{8}} \cdot \overset{10}{\cancel{80}} = 50$

4. (B) $6\% = .06 \qquad .06 \cdot b = .06b$

5. (D) $80 = .40x \qquad$ Divide by .40.

   $200 = x$

6. (B) $83\dfrac{1}{3}\% = \dfrac{5}{6} \qquad$ Multiply by 6. Divide by 5.

   $c = \dfrac{5}{6}x$

   $6c = 5x$

   $\dfrac{6c}{5} = x$

7. (E) $87\dfrac{1}{2}\% = \dfrac{7}{8} = \dfrac{14}{16}$

8. (D) $\dfrac{16}{40} = \dfrac{2}{5} = 40\%$

9. (B) $112\% = 1.12$

   $1.12 \cdot 80 = 89.6$

10. (C) $\dfrac{72}{60} = \dfrac{6}{5} = 120\%$

Exercise 1

1. (C) $3\dfrac{1}{2}\% = 3.5\% = .035$ To change a percent to a decimal, move the decimal point two places to the *left*.

2. (B) $85\% = \dfrac{85}{100} = \dfrac{17}{20}$

3. (E) To change a decimal to a percent, move the decimal point two places to the *right*.
   $4.6 = 460\%$

4. (C) $\dfrac{5}{\underset{3}{\cancel{12}}} \cdot \overset{25}{\cancel{100}} = \dfrac{125}{3} = 41\dfrac{2}{3}\%$

   To change a fraction to a percent, multiply by 100.

5. (B) $\dfrac{1}{2}\% = .5\% = .005$

Exercise 2

1. (C) $40\% - \dfrac{2}{5}$

   $\dfrac{2}{\cancel{5}} \cdot \overset{8}{\cancel{40}} = 16$

2. (E) $\begin{array}{r} 67 \\ \times\,42 \\ \hline 1\,34 \\ 26\,80 \\ \hline 28.14 \end{array}$

3. (A) $16\dfrac{2}{3}\% = \dfrac{1}{6}$

   $\dfrac{1}{\cancel{6}} \cdot \overset{20}{\cancel{120}} = 20$

4. (C) $\dfrac{1}{5}\% = .2\% = .002$

   $\begin{array}{r} 40 \\ \times\,.002 \\ \hline .0800 \end{array}$

5. (B) $r\% = \dfrac{r}{100}$

   $\dfrac{r}{100} \cdot s = \dfrac{rs}{100}$

Exercise 3

1. **(C)** $72 = .12x$
   $$7200 = 12x$$
   $$x = 600$$

2. **(D)** $80 = \dfrac{1}{8}x$
   $$640 = x$$

3. **(A)** $\dfrac{3}{8}x = 27$
   $$3x = 216$$
   $$x = 72$$

4. **(E)** $m = \dfrac{p}{100} \cdot x$
   $$100m = px$$
   $$\dfrac{100m}{p} = x$$

5. **(D)** $\dfrac{1}{2}x = r$
   $$x = 2r$$

Exercise 4

1. **(C)** $\dfrac{4}{80} = \dfrac{1}{20} \cdot \overset{5}{100} = 5\%$

2. **(B)** $\dfrac{1}{2}$ of 6 $= 3$

   $\dfrac{1}{4}$ of 60 $= 15$

   $\dfrac{3}{15} = \dfrac{1}{5}$ $= 20\%$

3. **(E)** $\dfrac{12}{96} = \dfrac{1}{8} = 12\dfrac{1}{2}\%$

4. **(C)** $\dfrac{48}{48} = 1 = 100\%$

5. **(D)** $\dfrac{x}{y} \cdot 100 = \dfrac{100x}{y}$

Exercise 5

1. **(A)** $36 = 1\dfrac{1}{2}x$
   $$36 = \dfrac{3}{2}x$$
   $$72 = 3x$$
   $$x = 24$$

2. **(D)** $300\% = 3$
   $$6 \cdot 3 = 18$$

3. **(B)** $\dfrac{120}{90} = \dfrac{4}{3} = 133\dfrac{1}{3}\%$

4. **(A)** $500 = 2x$
   $$250 = x$$

5. **(C)** $137.5\% = 1.375$

Retest

1. **(C)** $.25\% = \dfrac{.25}{100} = \dfrac{25}{10,000} = \dfrac{1}{400}$

2. **(D)** $\dfrac{3}{4}\% = .75\% = .0075$

3. **(E)** $12\% = .12$
   $.12 \cdot 80 = 9.6$

4. **(B)** $18 = .20x$ Divide by .20.
   $$90 = x$$

5. **(A)** $b\% = \dfrac{b}{100}$ $\dfrac{b}{\underset{50}{100}} \cdot \overset{3}{6} = \dfrac{3b}{50}$

6. **(B)** $62\dfrac{1}{2}\% = \dfrac{5}{8}$

   $m = \dfrac{5}{8}x$ Multiply by 8.

   $8m = 5x$ Divide by 5.

   $\dfrac{8m}{5} = x$

7. **(C)** $\dfrac{2}{12} = \dfrac{1}{6} = 16\dfrac{2}{3}\%$

8. **(E)** $140\% = 1.40$
   $1.40 \cdot 70 = 98$

9. **(C)** $280\% = \dfrac{280}{100} = \dfrac{28}{10} = \dfrac{14}{5}$

10. **(A)** $\dfrac{16}{12} = \dfrac{4}{3} = 133\dfrac{1}{3}\%$

Verbal Problems Involving Percent

6

DIAGNOSTIC TEST

Directions: Work out each problem. Circle the letter that appears before your answer.

Answers are at the end of the chapter.

1. A book dealer bought 100 books for $1250. If she sold 30% of these at $10 each and the rest at $15 each, what was her total profit?
   - (A) $350
   - (B) $1350
   - (C) $300
   - (D) $1050
   - (E) $100

2. The Fishman family income for one month is $2000. If 25% is spent for lodging, 35% for food, 5% for clothing, and 10% for savings, how many dollars are left for other expenses?
   - (A) $1500
   - (B) $400
   - (C) $500
   - (D) $1600
   - (E) $600

3. The enrollment of Kennedy High School dropped from 1200 to 1000 over a three-year period. What was the percent of decrease during this time?

   - (A) 20
   - (B) $16\frac{2}{3}$
   - (C) 25
   - (D) 200
   - (E) 2

4. A baseball team won 50 of the first 92 games played in a season. If the season consists of 152 games, how many more games must the team win to finish the season winning $62\frac{1}{2}$ % of games played?
   - (A) 37
   - (B) 45
   - (C) 40
   - (D) 95
   - (E) 19

5. The Strauss Insurance Company laid off 20% of its employees one year and then increased its staff by $12\frac{1}{2}$ % the following year. If the firm originally employed 120 workers, what was the net change in staff over the two-year period?
   - (A) Decrease of 12
   - (B) Increase of 15
   - (C) Decrease of 9
   - (D) Decrease of 24
   - (E) Increase of 12

6. How much money is saved by buying an article priced at $80 with a 40% discount, rather than buying an article marked at $90 with a discount of 35% then 10%?
   - (A) $4.65
   - (B) $1.50
   - (C) $10.50
   - (D) $3.15
   - (E) $4.25

7. In Central City, a property owner pays school taxes at the rate of 2% of the first $1500 of assessed valuation, 3% of the next $2000, 5% of the next $3000, and 6% of the remainder. How much must Mr. Williams pay in school taxes each year if his home is assessed at $8000?

    (A) $300
    (B) $230
    (C) $600
    (D) $330
    (E) $195

8. Jeffrey delivers newspapers for a salary of $20 per week plus a 4% commission on all sales. One week his sales amounted to $48. What was his income that week?

    (A) $19.20
    (B) $21.92
    (C) $1.92
    (D) $39.20
    (E) $32

9. At Baker High, 3 out of every 4 graduates go on to college. Of these, 2 out of every 3 graduate from college. What percent of students graduating from Baker High will graduate from college?

    (A) $66\frac{2}{3}$
    (B) 75
    (C) 50
    (D) $33\frac{1}{3}$
    (E) 25

10. The basic sticker price on Mr. Feldman's new car was $3200. The options he desired cost an additional $1800. What percent of the total price was made up of options?

    (A) $56\frac{1}{4}$
    (B) 36
    (C) 64
    (D) 18
    (E) 9

Certain types of business situations are excellent applications of percent. Study the examples on the following page carefully, as they are problems you will encounter in everyday life as well as on these examinations.

1. PERCENT OF INCREASE OR DECREASE

The percent of increase or decrease is found by putting the amount of increase or decrease over the original amount and changing this fraction to a percent by multiplying by 100.

Example:

> The number of automobiles sold by the Cadcoln Dealership increased from 300 one year to 400 the following year. What was the percent of increase?

Solution:

> There was an increase of 100, which must be compared to the original 300.
>
> $$\frac{100}{300} = \frac{1}{3} = 33\frac{1}{3}\%$$

Example:

> The Sunset School dismisses 20% of its staff of 150 due to budgetary problems. By what percent must it now increase its staff to return to the previous level?

Solution:

> $$20\% = \frac{1}{5} \qquad \frac{1}{5} \cdot 150 = 30$$
>
> The school now has 150 – 30 or 120 employees. To increase by 30, the percent of increase is
>
> $$\frac{30}{120} = \frac{1}{4} = 25\%.$$

Exercise 1

Work out each problem. Circle the letter that appears before your answer.

1. Mrs. Morris receives a salary raise from $25,000 to $27,500. Find the percent of increase.

   (A) 9
   (B) 10
   (C) 90
   (D) 15
   (E) 25

2. The population of Stormville has increased from 80,000 to 100,000 in the last twenty years. Find the percent of increase.

   (A) 20
   (B) 25
   (C) 80
   (D) 60
   (E) 10

3. The value of Super Company Stock dropped from $25 a share to $21 a share. Find the percent of decrease.

   (A) 4
   (B) 8
   (C) 12
   (D) 16
   (E) 20

4. The Rubins bought their home for $30,000 and sold it for $60,000. What was the percent of increase?

   (A) 100
   (B) 50
   (C) 200
   (D) 300
   (E) 150

5. During the pre-holiday rush, Martin's Department Store increased its sales staff from 150 to 200 persons. By what percent must it now decrease its sales staff to return to the usual number of salespersons?

   (A) 25
   (B) $33\frac{1}{3}$
   (C) 20
   (D) 40
   (E) 75

2. DISCOUNT

A discount is usually expressed as a percent of the marked price, which will be deducted from the marked price to determine the sale price. If an article is sold at a 20% discount, the buyer pays 80% of the marked price. Instead of first finding the amount of discount by finding 20% of the marked price and subtracting to find the sale price, it is shorter and easier to find 80% of the marked price directly.

Example:

A store offers a 25% discount on all appliances for paying cash. How much will a microwave oven marked at $400 cost if payment is made in cash?

Solution:

We can find 25% or $\frac{1}{4}$ of $400, which is $100, then subtract $100 from $400 to get a cash price of $300. The danger in this method is that the amount of discount, $100, is sure to be among the multiple-choice answers, as students often look for the first answer they get without bothering to finish the problem. It is safer, and easier, to realize that a 25% discount means 75% must be paid. 75% = $\frac{3}{4}$ and $\frac{3}{4}$ of $400 is $300.

Some problems deal with successive discounts. In such cases, the first discount is figured on the marked price, while the second discount is figured on the intermediate price.

Example:

Johnson's Hardware Store is having a moving sale in which everything in the store is being marked down 20% with an additional 5% discount for paying cash. What will be the net cost of a toaster, paid with cash, marked at $25?

Solution:

The first discount is 20% or $\frac{1}{5}$. We then pay $\frac{4}{5}$ of $25 or $20. An additional 5% is given off this amount. $\frac{5}{100} = \frac{1}{20}$ off. $\frac{19}{20} \cdot 20 = \$19$. The net price is $19.

Exercise 2

Work out each problem. Circle the letter that appears before your answer.

1. How much is saved by buying a freezer marked at $600 with a discount of 20% rather than one marked at $600 with a discount of 10% then 10%?

   (A) $6
   (B) $8
   (C) $10
   (D) $12
   (E) $20

2. Mr. Kaplan builds a home at a cost of $60,000. After pricing the home for sale by adding 25% of his expenses, he offers a discount of 20% to encourage sales. What did he make on the house?

   (A) $15,000
   (B) $1500
   (C) $0
   (D) $5000
   (E) $1200

3. Christmas cards are sold after Christmas for 90 cents a box instead of $1.20 a box. The rate of discount is

   (A) 20%
   (B) 25%
   (C) 30%
   (D) $33\frac{1}{3}\%$
   (E) 40%

4. A television set listed at $160 is offered at a $12\frac{1}{2}\%$ discount during a storewide sale. If an additional 3% is allowed on the net price for payment in cash, how much can Josh save by buying this set during the sale for cash?

   (A) $24.36
   (B) $24.80
   (C) $17.20
   (D) $24.20
   (E) $23.20

5. Pam pays $6 for a sweater after receiving a discount of 25%. What was the marked price of the sweater?

   (A) $9
   (B) $12
   (C) $7
   (D) $7.50
   (E) $8

3. COMMISSION

In order to inspire sales, many companies pay their salespeople a percentage of the money the salespeople bring in. This is called a commission.

Example:

Mr. Silver sells shoes at the Emporium, where he is paid $100 per week plus a 5% commission on all his sales. How much does he earn in a week in which his sales amount to $1840?

Solution:

Find 5% of $1840 and add this amount to $100.

$$\begin{array}{r} 1840 \\ \times\ .05 \\ \hline \$92.00 \end{array}\quad + \$100 = \$192$$

Example:

Audrey sells telephone order merchandise for a cosmetics company. She keeps 12% of all money collected. One month she was able to keep $108. How much did she forward to the cosmetics company?

Solution:

We must first find the total amount of her sales by asking: 108 is 12% of what number?

$108 = .12x$

$10800 = 12x$

$900 = x$

If Audrey collected $900 and kept $108, she sent the company $792.

Exercise 3

Work out each problem. Circle the letter that appears before your answer.

1. Janice receives a 6% commission for selling newspaper advertisements. If she sells 15 ads for $50 each, how much does she earn?

   (A) $30
   (B) $40
   (C) $45
   (D) $18
   (E) $450

2. Michael sells appliances and receives a salary of $125 per week plus a 5% commission on all sales over $750. How much does he earn in a week in which his sales amount to $2130?

   (A) $69
   (B) $294
   (C) $106.50
   (D) $194
   (E) $162.50

3. Mr. Rosen receives a salary of $100 per month plus a commission of 3% of his sales. What was the amount of his sales in a month in which he earned a total salary of $802?

   (A) $23,500
   (B) $23,400
   (C) $7800
   (D) $7900
   (E) $7700

4. Bobby sent $27 to the newspaper dealer for whom he delivers papers, after deducting his 10% commission. How many papers did he deliver if they sell for 20 cents each?

   (A) 150
   (B) 135
   (C) 600
   (D) 160
   (E) 540

5. Mrs. Mitherz wishes to sell her home. She must pay the real estate agent who makes the sale 8% of the selling price. At what price must she sell her home if she wishes to net $73,600?

   (A) $79,488
   (B) $75,000
   (C) $80,000
   (D) $82,400
   (E) $84,322

4. PROFIT AND LOSS

When a merchant purchases an item, he adds a percent of this cost to what he paid to arrive at a selling price. This amount is called his profit.

Example:

A radio sells for $40, giving the dealer a 25% profit. What was his cost?

Solution:

If the dealer gets back all of his cost plus an extra 25%, then the $40 sales price represents 125% of his cost.

$1.25x = 40$

$125x = 4000$

$x = \$32$

Example:

Joan's Boutique usually sells a handbag for $80, which yields a $33\frac{1}{3}$% profit. During a special sale, the profit is cut to 10%. What is the sale price of the handbag?

Solution:

$80 represents $133\frac{1}{3}$% of the cost.

$$\frac{4}{3}x = 80$$

$$4x = 240$$

$$x = 60$$

If the cost was $60 and the dealer wishes to add 10% for profit, he must add 10% of $60 or $6, making the sale price $66.

If a merchant sells an article for less than his cost, he takes a loss. A loss is figured as a percent of his cost in the same manner we figured a profit in the previous examples.

Exercise 4

Work out each problem. Circle the letter that appears before your answer.

1. Steve buys a ticket to the opera. At the last moment, he finds he cannot go and sells the ticket to Judy for $10, which was a loss of $16\frac{2}{3}$ %. What was the original price of the ticket?

   (A) $8.33
   (B) $16.66
   (C) $12
   (D) $11.66
   (E) $15

2. Alice bought a bicycle for $120. After using it for only a short time, she sold it to a bike store at a 20% loss. How much money did the bike store give Alice?

   (A) $24
   (B) $96
   (C) $144
   (D) $100
   (E) $108

3. Julie's Dress Shop sold a gown for $150, thereby making a 25% profit. What was the cost of the gown to the dress shop?

   (A) $120
   (B) $112.50
   (C) $117.50
   (D) $187.50
   (E) $125

4. If a music store sells a clarinet at a profit of 20% based on the selling price, what percent is made on the cost?

   (A) 20
   (B) 40
   (C) 25
   (D) 80
   (E) none of these

5. Radio House paid $60 for a tape player. At what price should it be offered for sale if the store offers customers a 10% discount but still wants to make a profit of 20% of the cost?

   (A) $64.80
   (B) $72
   (C) $79.20
   (D) $80
   (E) $84.20

5. TAXES

Taxes are a percent of money spent, money earned, or value.

Example:

Broome County has a 4% sales tax on appliances. How much will Mrs. Steinberg have to pay for a new dryer marked at $240?

Solution:

Find 4% of $240 to figure the tax and add this amount to $240. This can be done in one step by finding 104% of $240.

$$
\begin{array}{r}
240 \\
\times\,1.04 \\
\hline
960 \\
24000 \\
\hline
\$249.60
\end{array}
$$

Example:

The Social Security tax is $7\frac{1}{4}$ %. How much must Mrs. Grossman pay in a year if her salary is $2000 per month?

Solution:

Her annual salary is 12(2000) or $24,000. Find $7\frac{1}{4}$ % of $24,000.

$$
\begin{array}{r}
24{,}000 \\
\times\,.0725 \\
\hline
12\ 0000 \\
48\ 0000 \\
1680\ 0000 \\
\hline
\$1740.0000
\end{array}
$$

Exercise 5

Work out each problem. Circle the letter that appears before your answer.

1. In Manorville, the current rate for school taxes is 7.5% of property value. Find the tax on a house assessed at $20,000.

   (A) $150
   (B) $1500
   (C) $15,000
   (D) $1250
   (E) $105

2. The income tax in a certain state is figured at 2% of the first $1000, 3% of the next $2000, 4% of the next $3000, and 5% thereafter. Find the tax on an income of $25,000.

   (A) $1150
   (B) $1015
   (C) $295
   (D) $280
   (E) $187

3. The sales tax in Nassau County is 7%. If Mrs. Gutman paid a total of $53.50 for new curtains, what was the marked price of the curtains?

   (A) $49.75
   (B) $49
   (C) $57.25
   (D) $50
   (E) $45.86

4. Eric pays $r\%$ tax on an article marked at s dollars. How many dollars tax does he pay?

   (A) $\dfrac{s}{100r}$

   (B) rs

   (C) $\dfrac{100s}{r}$

   (D) $100rs$

   (E) $\dfrac{rs}{100}$

5. The sales tax on luxury items is 8%. If Mrs. Behr purchases a mink coat marked at $4000, what will be the total price for the coat, including tax?

   (A) $320
   (B) $4032
   (C) $4320
   (D) $4500
   (E) $500

RETEST

Work out each problem. Circle the letter that appears before your answer.

1. A TV sells for $121. What was the cost if the profit is 10% of the cost?
   (A) $110
   (B) $108.90
   (C) $120
   (D) $116
   (E) $111.11

2. Green's Sport Shop offers its salespeople an annual salary of $10,000 plus a 6% commission on all sales above $20,000. Every employee receives a Christmas bonus of $500. What are Mr. Cahn's total earnings in a year in which his sales amounted to $160,000?
   (A) $18,900
   (B) $18,400
   (C) $19,600
   (D) $20,100
   (E) $8900

3. A car dealer purchased 40 new cars at $6500 each. He sold 40% of them at $8000 each and the rest at $9000 each. What was his total profit?
   (A) $24,000
   (B) $60,000
   (C) $84,000
   (D) $344,000
   (E) $260,000

4. Mr. Adams' income rose from $20,000 one year to $23,000 the following year. What was the percent of increase?
   (A) 3%
   (B) 12%
   (C) 15%
   (D) 13%
   (E) 87%

5. The enrollment at Walden School is 1400. If 20% of the students study French, 25% study Spanish, 10% study Italian, 15% study German, and the rest study no language, how many students do not study a language, assuming each student may study only one language?
   (A) 30
   (B) 42
   (C) 560
   (D) 280
   (E) 420

6. How much money is saved by buying a car priced at $6000 with a single discount of 15% rather than buying the same car with a discount of 10% then 5%?
   (A) $51.30
   (B) $30
   (C) $780
   (D) $87
   (E) $900

7. At the Acme Cement Company, employees contribute to a welfare fund at the rate of 4% of the first $1000 earned, 3% of the next $1000, 2% of the next $1000, and 1% of any additional income. What will Mr. Morris contribute in a year in which he earns $20,000?
   (A) $290
   (B) $200
   (C) $90
   (D) $260
   (E) $240

8. A salesman receives a commission of $c\%$ on a sale of D dollars. Find his commission.
   (A) cD
   (B) $\dfrac{cD}{100}$
   (C) $100cD$
   (D) $\dfrac{c}{100D}$
   (E) $\dfrac{100c}{D}$

9. John buys a tape player for $54 after receiving a discount of 10%. What was the marked price?
   (A) $48.60
   (B) $59.40
   (C) $60
   (D) $61.40
   (E) $64

10. What single discount is equivalent to two successive discounts of 15% and 10%?
    (A) 25%
    (B) 24.5%
    (C) 24%
    (D) 23.5%
    (E) 23%

SOLUTIONS TO PRACTICE EXERCISES

Diagnostic Test

1. (E) $30\% = \dfrac{3}{10}$

   $\dfrac{3}{10} \cdot 100 = 30$ books at \$10 each

   $= \$300$ in sales

   $100 - 30 = 70$ books at \$15 each

   $= \$1050$ in sales

   Total sales $\$300 + \$1050 = \$1350$

   Total profit $\$1350 - \$1250 = \$100$

2. (C) $25\% + 35\% + 5\% + 10\% = 75\%$

   $100\% - 75\% = 25\%$ for other expenses

   $25\% = \dfrac{1}{4}$ $\dfrac{1}{4} \cdot \$2000 = \$500$

3. (B) Amount of decrease $= 200$

   Percent of decrease $= \dfrac{200}{1200} = \dfrac{1}{6} = 16\dfrac{2}{3}\%$

4. (B) $62\dfrac{1}{2}\% = \dfrac{5}{8}$

   $\dfrac{5}{8} \cdot 152 = 95$ total wins needed

   $95 - 50 = 45$ wins still needed

5. (A) $20\% = \dfrac{1}{5}$

   $\dfrac{1}{5} \cdot 120 = 24$ employees laid off

   New number of employees $= 96$

   $12\dfrac{1}{2}\% = \dfrac{1}{8}$

   $\dfrac{1}{8} \cdot 96 = 12$ employees added to staff

   Therefore, the final number of employees is 108. Net change is $120 - 108 =$ decrease of 12.

6. (A) $40\% = \dfrac{2}{5}$ $\dfrac{2}{5} \cdot 80 = \$32$ off

   $\$48$ net price

   $35\% = \dfrac{7}{20}$ $\dfrac{7}{20} \cdot 90 = \$31.50$ off

   $\$58.50$ net price

   $10\% = \dfrac{1}{10}$ $\dfrac{1}{10} \cdot 58.50 = \$5.85$ off

   $\$52.65$ net price

   $\$52.65 - \$48 = \$4.65$ was saved.

7. (D) 2% of \$1500 = \$30

   3% of \$2000 = \$60

   5% of \$3000 = \$150

   6% of (\$8000 − \$6500)

   = 6% of \$1500 = \$90

   Total tax = \$330

8. (B) He earns 4% of \$48.

   $\begin{array}{r} 48 \\ \times\,.04 \\ \hline \$1.92 \end{array}$

   Add this to his base salary of \$20: \$21.92.

9. (C) $\dfrac{\cancel{2}}{\cancel{3}} \cdot \dfrac{\cancel{3}}{\cancel{4}} = \dfrac{1}{2} = 50\%$ of the students will graduate from college.

10. (B) Total price is \$5000.

   Percent of total that was options $=$
   $\dfrac{1800}{5000} = \dfrac{9}{25} = 36\%$

Exercise 1

1. (B) Amount of increase = $2500

   Percent of increase = [amount of increase/original]

   $$\frac{25\cancel{00}}{250\cancel{00}} = \frac{1}{10} = 10\%$$

2. (B) Amount of increase = 20,000

   Percent of increase = $\frac{20,000}{80,000} = \frac{1}{4} = 25\%$

3. (D) Amount of decrease = $4

   Percent of decrease = $\frac{4}{25} = \frac{16}{100} = 16\%$

4. (A) Amount of increase $30,000

   Percent of increase = $\frac{30,000}{30,000} = 1 = 100\%$

5. (A) Amount of decrease = 50

   Percent of decrease = $\frac{50}{200} = \frac{1}{4} = 25\%$

Exercise 2

1. (A) $20\% = \frac{1}{5}$ $\frac{1}{5} \cdot 600 = \$120$ off

   $480 net price

   $10\% = \frac{1}{10}$ $\frac{1}{10} \cdot 600 = \$60$ off

   $540 first net price

   $\frac{1}{10} \cdot 540 = \$54$ off

   $486 net price

   Therefore, $6 is saved.

2. (C) $25\% = \frac{1}{4}$

   $\frac{1}{4} \cdot 60,000 = \$15,000$ added cost

   Original sale price = $75,000

   $20\% = \frac{1}{5}$ $\frac{1}{5} \cdot 75,000 = 15,000$ discount

   Final sale price $60,000

   Therefore he made nothing on the sale.

3. (B) Discount = 30 cents. Rate of discount is figured on the original price.

   $$\frac{30}{120} = \frac{1}{4} = 25\%$$

4. (D) $12\frac{1}{2}\% = \frac{1}{8}$ $\frac{1}{8} \cdot 160 = \$20$ discount

   New sale price = $140

   $3\% = \frac{3}{100}$ $\frac{3}{100} \cdot 140 = \frac{420}{100}$

   $= \$4.20$ second discount

   $135.80 final sale price

   Therefore, $160 – $135.80 or $24.20 was saved.

   Note: The amount saved is also the sum of the two discounts—$20 and $4.20.

5. (E) $6 is 75% of the marked price.

   $6 = \frac{3}{4}x$

   $24 = 3x$

   $x = \$8$

Exercise 3

1. (C) She sells 15 ads at $50 each for a total of $750. She earns 6% of this amount.

    $750 \times .06 = \$45.00$

2. (D) He earns 5% of ($2130 - $750).

    $1380 \times .05 = \$69.00$

    Add this to his base salary of $125: $194.

3. (B) If his base salary was $100, his commission amounted to $702. 702 is 3% of what?

    $702 = .03x$

    $70,200 = 3x$

    $\$23,400 = x$

4. (A) $27 is 90% of what he collected.

    $27 = .90x$

    $270 = 9x$

    $x = \$30$

    If each paper sells for 20 cents, he sold $\dfrac{30.00}{.20}$ or 150 papers.

5. (C) $73,600 is 92% of the selling price.

    $73,600 = .92x$

    $7,360,000 = 92x$

    $\$80,000 = x$

Exercise 4

1. (C) $16\dfrac{2}{3}\% = \dfrac{1}{6}$

    $10 is $\dfrac{5}{6}$ of the original price.

    $10 = \dfrac{5}{6}x$

    $60 = 5x$

    $x = 12$

2. (B) The store gave Alice 80% of the price she paid.

    $80\% = \dfrac{4}{5}$ $\dfrac{4}{5} \cdot 120 = \$96$

3. (A) $150 is 125% of the cost.

    $150 = 1.25x$

    $15,000 = 125x$

    $x = \$120$

4. (C) Work with an easy number such as $100 for the selling price.

    $20\% = \dfrac{1}{5}$ $\dfrac{1}{5} \cdot 100 = \$20$ profit, thereby making the cost $80. $\dfrac{20}{80} = \dfrac{1}{4} = 25\%$

5. (D) The dealer wishes to make 20% or $\dfrac{1}{5}$ of $60, which is $12 profit. The dealer wishes to clear $60 + $12 or $72. $72 will be 90% of the marked price.

    $72 = .90x$

    $720 = 9x$

    $x = \$80$

Exercise 5

1. **(B)**
$$\begin{array}{r} 20{,}000 \\ \times\ .075 \\ \hline 100\ 000 \\ 1400\ 000 \\ \hline 1500.000 \end{array}$$

2. **(A)**   2% of $1000 = $20
   3% of $2000 = $60
   4% of $3000 = $120
   5% of ($25,000 – $6,000)
   = 5% of $19,000 = $950

   Total tax = $1150

3. **(D)** $53.50 is 107% of the marked price

   $53.50 = 1.07x$
   $5350 = 107x$
   $x = \$50$

4. **(E)** $r\% = \dfrac{r}{100}$ $\dfrac{r}{100} \cdot s = \dfrac{rs}{100}$

5. **(C)**
$$\begin{array}{r} 4000 \\ \times\ .08 \\ \hline 320.00\ \text{tax} \end{array}$$ Total price $4320

Retest

1. **(A)** $121 is 110% of the cost.
   $121 = 1.10x$
   $1210 = 11x$
   $x = \$110$

2. **(A)**   He earns 6% of ($160,000 – $20,000).

$$\begin{array}{r} 140{,}000 \\ \times\ \ \ .06 \\ \hline \$8400.00 \end{array}$$

   Add this to his base salary of $10,000 and his Christmas bonus of $500: $18,900.

3. **(C)** $40\% = \dfrac{2}{5}$ $\dfrac{2}{5} \cdot 40 = 16$ cars at $8000 each = $128,000 in sales

   $40 – 16 = 24$ cars at $9000 each
   = $216,000 in sales

   Total sales: $128,000 + $216,000 = $344,000

   Total expense: $6500 \cdot 40 = $260,000

   Total profit: $344,000 – $260,000 = $84,000

4. **(C)**   Amount of increase = $3000

   Percent of increase $= \dfrac{3000}{20{,}000}$

   $= \dfrac{3}{20} = 15\%$

5. **(E)** 20% + 25% + 10% + 15% = 70%

   100% - 70% = 30% study no language

   $30\% = \dfrac{3}{10}$ $\dfrac{3}{10} \cdot 1400 = 420$

6. **(B)** $15\% = \dfrac{3}{20}$ $\dfrac{3}{20} \cdot \$6000 = \$900$ off

   $5100 net price

   $10\% = \dfrac{1}{10}$ $\dfrac{1}{10} \cdot \$6000 = \$600$ off

   $5400 first net price

   $5\% = \dfrac{1}{20}$ $\dfrac{1}{20} \cdot 5400 = \$270$ off

   $5130 net price

   $5130 – $5100 = $30 was saved.

7. **(D)** 4% of $1000 = $40

    3% of $1000 = $30

    2% of $1000 = $20

    1% of $17,000 = $170

    Total contribution = $260

8. **(B)** $c\% = \dfrac{c}{100}$ $\dfrac{c}{100} \cdot D = \dfrac{cD}{100}$

9. **(C)** $54 is 90% of the marked price.

    $54 = \dfrac{9}{10}x$

    $540 = 9x$

    $x = \$60$

10. **(D)** Work with an easy number such as $100.

    $15\% = \dfrac{3}{20}$ $\dfrac{3}{20} \cdot \$100 = \$15$ off

    $85 first net price

    $10\% = \dfrac{1}{10}$ $\dfrac{1}{10} \cdot \$85 = \$8.50$ off

    $76.50 net price

    $100 − $76.50 = $23.50 total discount

    $\dfrac{23.50}{100} = 23.5\%$

Averages

DIAGNOSTIC TEST

Directions: Work out each problem. Circle the letter that appears before your answer.

Answers are at the end of the chapter.

1. Find the average of the first ten positive even integers.
   - (A) 9
   - (B) 10
   - (C) 11
   - (D) 12
   - (E) $5\frac{1}{2}$

2. What is the average of $x - 4$, x, and $x + 4$?
   - (A) $3x$
   - (B) x
   - (C) $x - 1$
   - (D) $x + 1$
   - (E) $\frac{3x - 8}{3}$

3. Find the average of $\sqrt{.09}$, .4, and $\frac{1}{2}$.
   - (A) .31
   - (B) .35
   - (C) .04
   - (D) .4
   - (E) .45

4. Valerie received test grades of 93 and 88 on her first two French tests. What grade must she get on the third test to have an average of 92?
   - (A) 95
   - (B) 100
   - (C) 94
   - (D) 96
   - (E) 92

5. The average of W and another number is A. Find the other number.
   - (A) $A - W$
   - (B) $A + W$
   - (C) $\frac{1}{2}(A - W)$
   - (D) $\frac{1}{2}(A + W)$
   - (E) $2A - W$

6. The weight of three packages are 4 lb. 10 oz., 6 lb. 13 oz, and 3 lb. 6 oz. Find the average weight of these packages.
   - (A) 4 lb. 43 oz.
   - (B) 4 lb. $7\frac{1}{2}$ oz.
   - (C) 4 lb. 15 oz.
   - (D) 4 lb. 6 oz.
   - (E) 4 lb. 12 oz.

7. If Barbara drove for 4 hours at 50 miles per hour and then for 2 more hours at 60 miles per hour, what was her average rate, in miles per hour, for the entire trip?
   - (A) 55
   - (B) $53\frac{1}{3}$
   - (C) $56\frac{2}{3}$
   - (D) 53
   - (E) $54\frac{1}{2}$

8. Mr. Maron employs three secretaries at a salary of $140 per week and five salespeople at a salary of $300 per week. What is the average weekly salary paid to an employee?

(A) $55
(B) $190
(C) $240
(D) $200
(E) $185

9. Which of the following statements are always true?

I.   The average of any three consecutive even integers is the middle integer.
II. The average of any three consecutive odd integers is the middle integer.
III. The average of any three consecutive multiples of 5 is the middle number.

(A) I only
(B) II only
(C) I and II only
(D) I and III only
(E) I, II, and III

10. Mark has an average of 88 on his first four math tests. What grade must he earn on his fifth test in order to raise his average to 90?

(A) 92
(B) 94
(C) 96
(D) 98
(E) 100

1. SIMPLE AVERAGE

Most students are familiar with the method for finding an average and use this procedure frequently during the school year. To find the average of *n* numbers, find the sum of all the numbers and divide this sum by *n*.

Example:

Find the average of 12, 17, and 61.

Solution:

$$
\begin{array}{r}
12 \\
17 \\
+ 61 \\
\hline
3\overline{)90} \\
\hline
30
\end{array}
$$

When the numbers to be averaged form an evenly spaced series, the average is simply the middle number. If we are finding the average of an even number of terms, there will be no middle number. In this case, the average is halfway between the two middle numbers.

Example:

Find the average of the first 40 positive even integers.

Solution:

Since these 40 addends are evenly spaced, the average will be half way between the 20th and 21st even integers. The 20th even integer is 40 (use your fingers to count if needed) and the 21st is 42, so the average of the first 40 positive even integers that range from 2 to 80 is 41.

The above concept must be clearly understood as it would use up much too much time to add the 40 numbers and divide by 40. Using the method described, it is no harder to find the average of 100 evenly spaced terms than it is of 40 terms.

In finding averages, be sure the numbers being added are all of the same form or in terms of the same units. To average fractions and decimals, they must all be written as fractions or all as decimals.

Example:

Find the average of $87\frac{1}{2}\%$, $\frac{1}{4}$, and .6

Solution:

Rewrite each number as a decimal before adding.

$$
\begin{array}{r}
.875 \\
.25 \\
+ .6 \\
\hline
3\overline{)1.725} \\
\hline
.575
\end{array}
$$

Exercise 1

Work out each problem. Circle the letter that appears before your answer.

1. Find the average of $\sqrt{.49}$, $\frac{3}{4}$, and 80%.

   (A) .72
   (B) .75
   (C) .78
   (D) .075
   (E) .073

2. Find the average of the first 5 positive integers that end in 3.

   (A) 3
   (B) 13
   (C) 18
   (D) 23
   (E) 28

3. The five men on a basketball team weigh 160, 185, 210, 200, and 195 pounds. Find the average weight of these players.

   (A) 190
   (B) 192
   (C) 195
   (D) 198
   (E) 180

4. Find the average of a, $2a$, $3a$, $4a$, and $5a$.

   (A) $3a^5$
   (B) $3a$
   (C) $2.8a$
   (D) $2.8a^5$
   (E) 3

5. Find the average of $\frac{1}{2}$, $\frac{1}{3}$, and $\frac{1}{4}$.

   (A) $\frac{1}{9}$
   (B) $\frac{13}{36}$
   (C) $\frac{1}{27}$
   (D) $\frac{13}{12}$
   (E) $\frac{1}{3}$

2. TO FIND A MISSING NUMBER WHEN AN AVERAGE IS GIVEN

In solving this type of problem, it is easiest to use an algebraic equation that applies the definition of average. That is,

$$\text{average} = \frac{\text{sum of terms}}{\text{number of terms}}$$

Example:

The average of four numbers is 26. If three of the numbers are 50, 12, and 28, find the fourth number.

Solution:

$$\frac{50+12+28+x}{4} = 26$$
$$50+12+28+x = 104$$
$$90+x = 104$$
$$x = 14$$

An alternative method of solution is to realize that the number of units below 26 must balance the number of units above 26. 50 is 24 units *above* 26. 12 is 14 units *below* 26. 28 is 2 units *above* 26. Therefore, we presently have 26 units (24 + 2) *above* 26 and only 14 units *below* 26. Therefore the missing number must be 12 units *below* 26, making it 14. When the numbers are easy to work with, this method is usually the fastest. Just watch your arithmetic.

Exercise 2

Work out each problem. Circle the letter that appears before your answer.

1. Dick's average for his freshman year was 88, his sophomore year was 94, and his junior year was 91. What average must he have in his senior year to leave high school with an average of 92?

   (A) 92
   (B) 93
   (C) 94
   (D) 95
   (E) 96

2. The average of X, Y, and another number is M. Find the missing number.

   (A) $3M - X + Y$
   (B) $3M - X - Y$
   (C) $\dfrac{M + X + Y}{3}$
   (D) $M - X - Y$
   (E) $M - X + Y$

3. The average of two numbers is $2x$. If one of the numbers is $x + 3$, find the other number.

   (A) $x - 3$
   (B) $2x - 3$
   (C) $3x - 3$
   (D) -3
   (E) $3x + 3$

4. On consecutive days, the high temperature in Great Neck was 86°, 82°, 90°, 92°, 80°, and 81°. What was the high temperature on the seventh day if the average high for the week was 84°?

   (A) 79°
   (B) 85°
   (C) 81°
   (D) 77°
   (E) 76°

5. If the average of five consecutive integers is 17, find the largest of these integers.

   (A) 17
   (B) 18
   (C) 19
   (D) 20
   (E) 21

3. WEIGHTED AVERAGE

When some numbers among terms to be averaged occur more than once, they must be given the appropriate weight. For example, if a student received four grades of 80 and one of 90, his average would not be the average of 80 and 90, but rather the average of 80, 80, 80, 80, and 90.

Example:

Mr. Martin drove for 6 hours at an average rate of 50 miles per hour and for 2 hours at an average rate of 60 miles per hour. Find his average rate for the entire trip.

Solution:

$$\frac{6(50)+2(60)}{8} = \frac{300+120}{8} = \frac{420}{8} = 52\frac{1}{2}$$

Since he drove many more hours at 50 miles per hour than at 60 miles per hour, his average rate should be closer to 50 than to 60, which it is. In general, average rate can always be found by dividing the total distance covered by the total time spent traveling.

Exercise 3

Work out each problem. Circle the letter that appears before your answer.

1. In a certain gym class, 6 girls weigh 120 pounds each, 8 girls weigh 125 pounds each, and 10 girls weigh 116 pounds each. What is the average weight of these girls?

   (A) 120
   (B) 118
   (C) 121
   (D) 122
   (E) 119

2. In driving from San Francisco to Los Angeles, Arthur drove for three hours at 60 miles per hour and for 4 hours at 55 miles per hour. What was his average rate, in miles per hour, for the entire trip?

   (A) 57.5
   (B) 56.9
   (C) 57.1
   (D) 58.2
   (E) 57.8

3. In the Linwood School, five teachers earn $15,000 per year, three teachers earn $17,000 per year, and one teacher earns $18,000 per year. Find the average yearly salary of these teachers.

   (A) $16,667
   (B) $16,000
   (C) $17,000
   (D) $16,448
   (E) $16,025

4. During the first four weeks of summer vacation, Danny worked at a camp earning $50 per week. During the remaining six weeks of vacation, he worked as a stock boy earning $100 per week. What was his average weekly wage for the summer?

   (A) $80
   (B) $75
   (C) $87.50
   (D) $83.33
   (E) $82

5. If M students each received a grade of P on a physics test and N students each received a grade of Q, what was the average grade for this group of students?

   (A) $\dfrac{P+Q}{M+N}$

   (B) $\dfrac{PQ}{M+N}$

   (C) $\dfrac{MP+NQ}{M+N}$

   (D) $\dfrac{MP+NQ}{P+Q}$

   (E) $\dfrac{M+N}{P+Q}$

RETEST

Work out each problem. Circle the letter that appears before your answer.

1. Find the average of the first 14 positive odd integers.

   (A) 7.5
   (B) 13
   (C) 14
   (D) 15
   (E) 14.5

2. What is the average of $2x - 3$, $x + 1$, and $3x + 8$?

   (A) $6x + 6$
   (B) $2x - 2$
   (C) $2x + 4$
   (D) $2x + 2$
   (E) $2x - 4$

3. Find the average of $\frac{1}{5}$, 25%, and .09.

   (A) $\frac{2}{3}$
   (B) .18
   (C) .32
   (D) 20%
   (E) $\frac{1}{4}$

4. Andy received test grades of 75, 82, and 70 on three French tests. What grade must he earn on the fourth test to have an average of 80 on these four tests?

   (A) 90
   (B) 93
   (C) 94
   (D) 89
   (E) 96

5. The average of $2P$, $3Q$, and another number is S. Represent the third number in terms of P, Q, and S.

   (A) $S - 2P - 3Q$
   (B) $S - 2P + 3Q$
   (C) $3S - 2P + 3Q$
   (D) $3S - 2P - 3Q$
   (E) $S + 2P - 3Q$

6. The students of South High spent a day on the street collecting money to help cure birth defects. In counting up the collections, they found that 10 cans contained $5.00 each, 14 cans contained $6.50 each, and 6 cans contained $7.80 each. Find the average amount contained in each of these cans.

   (A) $6.14
   (B) $7.20
   (C) $6.26
   (D) $6.43
   (E) $5.82

7. The heights of the five starters on the Princeton basketball team are 6′ 6″, 6′ 7″, 6′ 9″, 6′ 11″, and 7′. Find the average height of these men.

   (A) $6'\,8\frac{1}{5}''$
   (B) $6'\,9''$
   (C) $6'\,9\frac{3}{5}''$
   (D) $6'\,9\frac{1}{5}''$
   (E) $6'\,9\frac{1}{2}''$

8. Which of the following statements is always true?

   I. The average of the first twenty odd integers is 10.5
   II. The average of the first ten positive integers is 5.
   III. The average of the first 4 positive integers that end in 2 is 17.

   (A) I only
   (B) II only
   (C) III only
   (D) I and III only
   (E) I, II, and III

9. Karen drove 40 miles into the country at 40 miles per hour and returned home by bus at 20 miles per hour. What was her average rate in miles per hour for the round trip?

    (A) 30

    (B) $25\frac{1}{2}$

    (C) $26\frac{2}{3}$

    (D) 20

    (E) $27\frac{1}{3}$

10. Mindy's average monthly salary for the first four months she worked was $300. What must be her average monthly salary for each of the next 8 months so that her average monthly salary for the year is $350?

    (A) $400
    (B) $380
    (C) $390
    (D) $375
    (E) $370

SOLUTIONS TO PRACTICE EXERCISES

Diagnostic Test

1. **(C)** The integers are 2, 4, 6, 8, 10, 12, 14, 16, 18, 20. Since these are evenly spaced, the average is the average of the two middle numbers, 10 and 12, or 11.

2. **(B)** These numbers are evenly spaced, so the average is the middle number x.

3. **(D)** $\sqrt{.09} = .3$

   $\dfrac{1}{2} = .5$

   $.4 = .4$

   $3)\underline{1.2}$
   $\quad\ .4$

4. **(A)** 93 is 1 above 92; 88 is 4 below 92. So far, she has 1 point above 92 and 4 points below 92. Therefore, she needs another 3 points above 92, making a required grade of 95.

5. **(E)** $\dfrac{W + x}{2} = A$

   $W + x = 2A$

   $x = 2A - W$

6. **(C)** 4 lb. 10 oz.

   6 lb. 13 oz.

   $+\ \underline{\ 3\ \text{lb.}\ 6\ \text{oz.}}$

   13 lb. 29 oz.

   $\dfrac{13\text{lb. }29\text{oz.}}{3} = \dfrac{12\text{lb. }45\text{oz.}}{3} = 4\text{ lb. 15 oz.}$

7. **(B)** $4(50) = 200$

   $2(60) = \underline{120}$

   $\qquad 6)\overline{320}$

   $\qquad\quad 53\dfrac{1}{3}$

8. **(C)** $3(140) = \quad 420$

   $5(300) = \underline{1500}$

   $\qquad 8)\overline{1920}$

   $\qquad\quad 240$

9. **(E)** The average of any three numbers that are evenly spaced is the middle number.

10. **(D)** Since 88 is 2 below 90, Mark is 8 points below 90 after the first four tests. Thus, he needs a 98 to make the required average of 90.

Exercise 1

1. **(B)** $\sqrt{.49} = .7$

   $\dfrac{3}{4} = .75$

   $80\% = \underline{.80}$

   $3)\overline{2.25}$
   $\quad\ .75$

2. **(D)** The integers are 3, 13, 23, 33, 43. Since these are evenly spaced, the average is the middle integer, 23.

3. **(A)** $160 + 185 + 210 + 200 + 195 = 950$

   $\dfrac{950}{5} = 190$

4. **(B)** These numbers are evenly spaced, so the average is the middle number, $3a$.

5. **(B)** $\dfrac{1}{2} + \dfrac{1}{3} + \dfrac{1}{4} = \dfrac{6}{12} + \dfrac{4}{12} + \dfrac{3}{12} = \dfrac{13}{12}$

   To divide this sum by 3, multiply by $\dfrac{1}{3}$

   $\dfrac{13}{12} \cdot \dfrac{1}{3} = \dfrac{13}{36}$

Exercise 2

1. **(D)** 88 is 4 below 92; 94 is 2 above 92; 91 is 1 below 92. So far, he has 5 points below 92 and only 2 above. Therefore, he needs another 3 points above 92, making the required grade 95.

2. **(B)** $\dfrac{X+Y+x}{3} = M$

   $X + Y + x = 3M$

   $x = 3M - X - Y$

3. **(C)** $\dfrac{(x+3)+n}{2} = 2x$

   $x + 3 + n = 4x$

   $n = 3x - 3$

4. **(D)** 86° is 2 above the average of 84; 82° is 2 below; 90° is 6 above; 92° is 8 above; 80° is 4 below; and 81° is 3 below. So far, there are 16° above and 9° below. Therefore, the missing term is 7° below the average, or 77°.

5. **(C)** 17 must be the middle integer, since the five integers are consecutive and the average is, therefore, the middle number. The numbers are 15, 16, 17, 18, and 19.

Exercise 3

1. **(A)**
   $6(120) = 720$
   $8(125) = 1000$
   $10(116) = \underline{1160}$

   $24\overline{)2880}$
   120

2. **(C)**
   $3(60) = 180$
   $4(55) = \underline{220}$

   $7\overline{)400}$

   $57\dfrac{1}{7},$

   which is 57.1 to the nearest tenth.

3. **(B)**
   $5(15,000) = 75,000$
   $3(17,000) = 51,000$
   $1(18,000) = \underline{18,000}$

   $9\overline{)144,000}$
   $16,000$

4. **(A)**
   $4(50) = 200$
   $6(100) = \underline{600}$

   $10\overline{)800}$
   80

5. **(C)**
   $M(P) = MP$
   $N(Q) = NQ$

   $\overline{}$
   $MP + NQ$

   Divide by the number of students, $M + N$.

Retest

1. **(C)** The integers are 1, 3, 5, 7, 9, 11, 13, 15, 17, 19, 21, 23, 25, 27. Since these are evenly spaced, the average is the average of the two middle numbers 13 and 15, or 14.

2. **(D)**

$$2x - 3$$
$$x + 1$$
$$\underline{+3x + 8}$$
$$6x + 6$$

$$\frac{6x + 6}{3} = 2x + 2$$

3. **(B)**

$$\frac{1}{5} = .20$$
$$25\% = .25$$
$$.09 = \underline{.09}$$
$$3\overline{).54}$$
$$.18$$

4. **(B)** 75 is 5 below 80; 82 is 2 above 80, 70 is 10 below 80. So far, he is 15 points below and 2 points above 80. Therefore, he needs another 13 points above 80, or 93.

5. **(D)**

$$\frac{2P + 3Q + x}{3} = S$$
$$2P + 3Q + x = 3S$$
$$x = 3S - 2P - 3Q$$

6. **(C)**

$$10(\$5.00) = \$50$$
$$14(\$6.50) = \$91$$
$$6(\$7.80) = \$46.80$$
$$30\overline{)\$187.80}$$
$$\$6.26$$

7. **(B)** $6'6'' + 6'7'' + 6'11'' + 6'9'' + 7' = 31'33'' = 33'9''$

$$\frac{33'9''}{5} = 6'9''$$

8. **(C)** I. The average of the first twenty *positive* integers is 10.5.

II. The average of the first ten positive integers is 5.5.

III. The first four positive integers that end in 2 are 2, 12, 22, and 32. Their average is 17.

9. **(C)** Karen drove for 1 hour into the country and returned home by bus in 2 hours. Since the total distance traveled was 80 miles, her average rate for the round trip was $\frac{80}{3}$ or $26\frac{2}{3}$ miles per hour.

10. **(D)** Since $300 is $50 below $350, Mindy's salary for the first four months is $200 below $350. Therefore, her salary for each of the next 8 months must be $\frac{\$200}{8}$ or $25 above the average of $350, thus making the required salary $375.

Concepts of Algebra—Signed Numbers and Equations

8

DIAGNOSTIC TEST

Directions: Work out each problem. Circle the letter that appears before your answer.

Answers are at the end of the chapter.

1. When +4 is added to –6, the sum is

    (A) –10
    (B) +10
    (C) –24
    (D) –2
    (E) +2

2. The product of $(-3)(+4)\left(-\dfrac{1}{2}\right)\left(-\dfrac{1}{3}\right)$ is

    (A) –1
    (B) –2
    (C) +2
    (D) –6
    (E) +6

3. When the product of (-12) and $\left(+\dfrac{1}{4}\right)$ is

    divided by the product of (-18) and $\left(-\dfrac{1}{3}\right)$, the

    quotient is

    (A) +2
    (B) –2
    (C) $+\dfrac{1}{2}$
    (D) $-\dfrac{1}{2}$
    (E) $-\dfrac{2}{3}$

4. Solve for x: $ax + b = cx + d$

    (A) $\dfrac{d-b}{ac}$
    (B) $\dfrac{d-b}{a+c}$
    (C) $\dfrac{d-b}{a-c}$
    (D) $\dfrac{b-d}{ac}$
    (E) $\dfrac{b-d}{a-c}$

5. Solve for y: $7x - 2y = 2$
    $3x + 4y = 30$

    (A) 2
    (B) 6
    (C) 1
    (D) 11
    (E) –4

6. Solve for x: $x + y = a$
    $x - y = b$

    (A) $a + b$
    (B) $a - b$
    (C) $\dfrac{1}{2}(a + b)$
    (D) $\dfrac{1}{2}ab$
    (E) $\dfrac{1}{2}(a - b)$

7. Solve for x: $4x^2 - 2x = 0$

    (A) $\frac{1}{2}$ only

    (B) 0 only

    (C) $-\frac{1}{2}$ only

    (D) $\frac{1}{2}$ or 0

    (E) $-\frac{1}{2}$ or 0

8. Solve for x: $x^2 - 4x - 21 = 0$

    (A) 7 or 3
    (B) −7 or −3
    (C) −7 or 3
    (D) 7 or −3
    (E) none of these

9. Solve for x: $\sqrt{x+1} - 3 = -7$

    (A) 15
    (B) 47
    (C) 51
    (D) 39
    (E) no solution

10. Solve for x: $\sqrt{x^2 + 7} - 1 = x$

    (A) 9
    (B) 3
    (c) −3
    (D) 2
    (E) no solution

1. SIGNED NUMBERS

The rules for operations with signed numbers are basic to successful work in algebra. Be sure you know, and can apply, the following rules.

Addition: To add numbers with the same sign, add the magnitudes of the numbers and keep the same sign. To add numbers with different signs, subtract the magnitudes of the numbers and use the sign of the number with the greater magnitude.

Example:

Add the following:

+4	−4	−4	+4
+7	−7	+7	−7
+11	−11	+3	−3

Subtraction: Change the sign of the number to be subtracted and proceed with the rules for addition. Remember that subtracting is really adding the additive inverse.

Example:

Subtract the following:

+4	−4	−4	+4
+7	−7	+7	−7
−3	+3	−11	+11

Multiplication: If there is an odd number of negative factors, the product is negative. An even number of negative factors gives a positive product.

Example:

Find the following products:

$$(+4)(+7) = +28 \qquad (-4)(-7) = +28$$
$$(+4)(-7) = -28 \qquad (-4)(+7) = -28$$

Division: If the signs are the same, the quotient is positive. If the signs are different, the quotient is negative.

Example:

Divide the following:

$$\frac{+28}{+4} = +7 \qquad \frac{-28}{+4} = -7$$

$$\frac{-28}{-4} = +7 \qquad \frac{+28}{-4} = -7$$

Exercise 1

Work out each problem. Circle the letter that appears before your answer.

1. At 8 a.m. the temperature was –4°. If the temperature rose 7 degrees during the next hour, what was the thermometer reading at 9 a.m.?

    (A) +11°
    (B) –11°
    (C) +7°
    (D) +3°
    (E) –3°

2. In Asia, the highest point is Mount Everest, with an altitude of 29,002 feet, while the lowest point is the Dead Sea, 1286 feet below sea level. What is the difference in their elevations?

    (A) 27,716 feet
    (B) 30,288 feet
    (C) 28,284 feet
    (D) 30,198 feet
    (E) 27,284 feet

3. Find the product of (–6)(–4)(–4) and (–2).

    (A) –16
    (B) +16
    (C) –192
    (D) +192
    (E) –98

4. The temperatures reported at hour intervals on a winter evening were +4°, 0°, –1°, –5°, and –8°. Find the average temperature for these hours.

    (A) –10°
    (B) –2°
    (C) +2°
    (D) $-2\frac{1}{2}$°
    (E) –3°

5. Evaluate the expression $5a - 4x - 3y$ if $a = -2$, $x = -10$, and $y = 5$.

    (A) +15
    (B) +25
    (C) –65
    (D) –35
    (E) +35

2. SOLUTION OF LINEAR EQUATIONS

Equations are the basic tools of algebra. The techniques of solving an equation are not difficult. Whether an equation involves numbers or only letters, the basic steps are the same.

1. If there are fractions or decimals, remove them by multiplication.

2. Remove any parentheses by using the distributive law.

3. Collect all terms containing the unknown for which you are solving on the same side of the equal sign. Remember that whenever a term crosses the equal sign from one side of the equation to the other, it must pay a toll. That is, it must change its sign.

4. Determine the coefficient of the unknown by combining similar terms or factoring when terms cannot be combined.

5. Divide both sides of the equation by the coefficient.

Example:

Solve for x: $5x - 3 = 3x + 5$

Solution:

$2x = 8$
$x = 4$

Example:

Solve for x: $\dfrac{2}{3}x - 10 = \dfrac{1}{4}x + 15$

Solution:

Multiply by 12. $8x - 120 = 3x + 180$
$5x = 300$
$x = 60$

Example:

Solve for x: $.3x + .15 = 1.65$

Solution:

Multiply by 100. $30x + 15 = 165$
$30x = 150$
$x = 5$

Example:

Solve for x: $ax - r = bx - s$

Solution:

$ax - bx = r - s$
$x(a - b) = r - s$
$x = \dfrac{r - s}{a - b}$

Example:

Solve for x: $6x - 2 = 8(x - 2)$

Solution:

$6x - 2 = 8x - 16$
$14 = 2x$
$x = 7$

Exercise 2

Work out each problem. Circle the letter that appears before your answer.

1. Solve for x: $3x - 2 = 3 + 2x$

   (A) 1
   (B) 5
   (C) −1
   (D) 6
   (E) −5

2. Solve for a: $8 - 4(a - 1) = 2 + 3(4 - a)$

   (A) $-\dfrac{5}{3}$

   (B) $-\dfrac{7}{3}$

   (C) 1
   (D) −2
   (E) 2

3. Solve for y: $\dfrac{1}{8}y + 6 = \dfrac{1}{4}y$

   (A) 48
   (B) 14
   (C) 6
   (D) 1
   (E) 2

4. Solve for x: $.02(x - 2) = 1$

   (A) 2.5
   (B) 52
   (C) 1.5
   (D) 51
   (E) 6

5. Solve for x: $4(x - r) = 2x + 10r$

   (A) $7r$
   (B) $3r$
   (C) r
   (D) $5.5r$
   (E) $2\dfrac{1}{3}r$

3. SIMULTANEOUS EQUATIONS IN TWO UNKNOWNS

In solving equations with two unknowns, it is necessary to work with two equations simultaneously. The object is to eliminate one of the unknowns, resulting in an equation with one unknown that can be solved by the methods of the previous section. This can be done by multiplying one or both equations by suitable constants in order to make the coefficients of one of the unknowns the same. Remember that multiplying *all* terms in an equation by the same constant does not change its value. The unknown can then be removed by adding or subtracting the two equations. When working with simultaneous equations, always be sure to have the terms containing the unknowns on one side of the equation and the remaining terms on the other side.

Example:

Solve for x: $7x + 5y = 15$
$5x - 9y = 17$

Solution:

Since we wish to solve for x, we would like to eliminate the y terms. This can be done by multiplying the top equation by 9 and the bottom equation by 5. In doing this, both y coefficients will have the same magnitude.

Multiplying the first by 9, we have

$63x + 45y = 135$

Multiplying the second by 5, we have

$25x - 45y = 85$

Since the y terms now have opposite signs, we can eliminate y by adding the two equations. If they had the same signs, we would eliminate by subtracting the two equations.

Adding, we have

$$63x + 45y = 135$$
$$\underline{25x - 45y = 85}$$
$$88x \qquad = 220$$

$$x = \frac{220}{88} = 2\frac{1}{2}$$

Since we were only asked to solve for x, we stop here. If we were asked to solve for both x and y, we would now substitute $2\frac{1}{2}$ for x in either equation and solve the resulting equation for y.

$7(2.5) + 5y = 15$
$17.5 + 5y = 15$
$5y = -2.5$
$y = -.5 \text{ or } -\frac{1}{2}$

Example:

Solve for x: $ax + by = r$
$cx - dy = s$

Solution:

Multiply the first equation by d and the second by b to eliminate the y terms by addition.
$adx + bdy = dr$
$\underline{bcx - bdy = bs}$
$adx + bcx = dr + bs$
Factor out x to determine the coefficient of x.
$x(ad + bc) = dr + bs$
$$x = \frac{dr + bs}{ad + bc}$$

Exercise 3

Work out each problem. Circle the letter that appears before your answer.

1. Solve for x: $x - 3y = 3$
    $$2x + 9y = 11$$

    (A) 2
    (B) 3
    (C) 4
    (D) 5
    (E) 6

2. Solve for x: $.6x + .2y = 2.2$

    $$.5x - .2y = 1.1$$

    (A) 1
    (B) 3
    (C) 30
    (D) 10
    (E) 11

3. Solve for y: $2x + 3y = 12b$
    $$3x - y = 7b$$

    (A) $7\frac{1}{7}b$
    (B) $2b$
    (C) $3b$
    (D) $1\frac{2}{7}$
    (E) $-b$

4. If $2x = 3y$ and $5x + y = 34$, find y.

    (A) 4
    (B) 5
    (C) 6
    (D) 6.5
    (E) 10

5. If $x + y = -1$ and $x - y = 3$, find y.

    (A) 1
    (B) −2
    (C) −1
    (D) 2
    (E) 0

4. QUADRATIC EQUATIONS

In solving quadratic equations, there will always be two roots, even though these roots may be equal. A complete quadratic equation is of the form $ax^2 + bx + c = 0$, where a, b, and c are integers. At the level of this examination, $ax^2 + bx + c$ can always be factored. If b and/or c is equal to 0, we have an incomplete quadratic equation, which can still be solved by factoring and will still have two roots.

Example:

> $x^2 + 5x = 0$

Solution:

> Factor out a common factor of x.
> $x(x + 5) = 0$
> If the product of two factors is 0, either factor may be set equal to 0, giving $x = 0$ or $x + 5 = 0$. From these two linear equations, we find the two roots of the given quadratic equation to be $x = 0$ and $x = -5$.

Example:

> $6x^2 - 8x = 0$

Solution:

> Factor out a common factor of $2x$.
> $2x(3x - 4) = 0$
> Set each factor equal to 0 and solve the resulting linear equations for x.
> $2x = 0$ $3x - 4 = 0$
> $x = 0$ $3x = 4$
> $x = \dfrac{4}{3}$
> The roots of the given quadratic are 0 and $\dfrac{4}{3}$.

Example:

> $x^2 - 9 = 0$

Solution:

> $x^2 = 9$
> $x = \pm 3$
> Remember there must be two roots. This equation could also have been solved by factoring $x^2 - 9$ into $(x + 3)(x - 3)$ and setting each factor equal to 0. Remember that the difference of two perfect squares can always be factored, with one factor being the sum of the two square roots and the second being the difference of the two square roots.

Example:

> $x^2 - 8 = 0$

Solution:

> Since 8 is not a perfect square, this cannot be solved by factoring.
> $x^2 = 8$
> $x = \pm\sqrt{8}$
> Simplifying the radical, we have $\sqrt{4} \cdot \sqrt{2}$, or $x = \pm 2\sqrt{2}$

Example:

$$16x^2 - 25 = 0$$

Solution:

Factoring, we have $\qquad (4x - 5)(4x + 5) = 0$

Setting each factor equal to 0, we have $\qquad x = \pm \dfrac{5}{4}$

If we had solved without factoring, we would have found $\quad 16x^2 = 25$

$$x^2 = \frac{25}{16}$$

$$x = \pm \frac{5}{4}$$

Example:

$$x^2 + 6x + 8 = 0$$

Solution:

$(x + 2)(x + 4) = 0$

If the last term of the trinomial is positive, both binomial factors must have the same sign, since the last two terms multiply to a positive product. If the middle term is also positive, both factors must be positive since they also add to a positive sum. Setting each factor equal to 0, we have $x = -4$ or $x = -2$

Example:

$$x^2 - 2x - 15 = 0$$

Solution:

We are now looking for two numbers that multiply to -15; therefore they must have opposite signs. To give -2 as a middle coefficient, the numbers must be -5 and $+3$.
$(x - 5)(x + 3) = 0$
This equation gives the roots 5 and -3.

Exercise 4

Work out each problem. Circle the letter that appears before your answer.

1. Solve for x: $x^2 - 8x - 20 = 0$

    (A) 5 and -4
    (B) 10 and -2
    (C) -5 and 4
    (D) -10 and -2
    (E) -10 and 2

2. Solve for x: $25x^2 - 4 = 0$

    (A) $\dfrac{4}{25}$ and $-\dfrac{4}{25}$
    (B) $\dfrac{2}{5}$ and $-\dfrac{2}{5}$
    (C) $\dfrac{2}{5}$ only
    (D) $-\dfrac{2}{5}$ only
    (E) none of these

3. Solve for x: $6x^2 - 42x = 0$

    (A) 7 only
    (B) -7 only
    (C) 0 only
    (D) 7 and 0
    (E) -7 and 0

4. Solve for x: $x^2 - 19x + 48 = 0$

    (A) 8 and 6
    (B) 24 and 2
    (C) -16 and -3
    (D) 12 and 4
    (E) none of these

5. Solve for x: $3x^2 = 81$

    (A) $9\sqrt{3}$
    (B) $\pm 9\sqrt{3}$
    (C) $3\sqrt{3}$
    (D) $\pm 3\sqrt{3}$
    (E) ± 9

5. EQUATIONS CONTAINING RADICALS

In solving equations containing radicals, it is important to get the radical alone on one side of the equation. Then square both sides to eliminate the radical sign. Solve the resulting equation. Remember that all solutions to radical equations must be checked, as squaring both sides may sometimes result in extraneous roots. In squaring each side of an equation, do not make the mistake of simply squaring each term. The entire side of the equation must be multiplied by itself.

Example:

$$\sqrt{x-3} = 4$$

Solution:

$$x - 3 = 16$$
$$x = 19$$

Checking, we have $\sqrt{16} = 4$, which is true.

Example:

$$\sqrt{x-3} = -4$$

Solution:

$$x - 3 = 16$$
$$x = 19$$

Checking, we have $\sqrt{16} = -4$, which is not true, since the radical sign means the principal, or positive, square root only. is 4, not –4; therefore, this equation has no solution.

Example:

$$\sqrt{x^2 - 7} + 1 = x$$

Solution:

First get the radical alone on one side, then square.

$$\sqrt{x^2 - 7} = x - 1$$
$$x^2 - 7 = x^2 - 2x + 1$$
$$-7 = -2x + 1$$
$$2x = 8$$
$$x = 4$$

Checking, we have $\sqrt{9} + 1 = 4$
$$3 + 1 = 4,$$

which is true.

Exercise 5

Work out each problem. Circle the letter that appears before your answer.

1. Solve for y: $\sqrt{2y} + 11 = 15$

    (A) 4
    (B) 2
    (C) 8
    (D) 1
    (E) no solution

2. Solve for x: $4\sqrt{2x-1} = 12$

    (A) 18.5
    (B) 4
    (C) 10
    (D) 5
    (E) no solution

3. Solve for x: $\sqrt{x^2 - 35} = 5 - x$

    (A) 6
    (B) –6
    (C) 3
    (D) –3
    (E) no solution

4. Solve for y: $26 = 3\sqrt{2y} + 8$

    (A) 6
    (B) 18
    (C) 3
    (D) –6
    (E) no solution

5. Solve for x: $\sqrt{\dfrac{2x}{5}} = 4$

    (A) 10
    (B) 20
    (C) 30
    (D) 40
    (E) no solution

RETEST

Work out each problem. Circle the letter that appears before your answer.

1. When –5 is subtracted from the sum of –3 and +7, the result is

   (A) +15
   (B) –1
   (C) –9
   (D) +9
   (E) +1

2. The product of $\left(-\frac{1}{2}\right)(-4)(+12)\left(-\frac{1}{6}\right)$ is

   (A) 2
   (B) –2
   (C) 4
   (D) –4
   (E) –12

3. When the sum of –4 and –5 is divided by the product of 9 and $-\frac{1}{27}$, the result is

   (A) –3
   (B) +3
   (C) –27
   (D) +27
   (E) $-\frac{1}{3}$

4. Solve for x: $7b + 5d = 5x - 3b$

   (A) $2bd$
   (B) $2b + d$
   (C) $5b + d$
   (D) $3bd$
   (E) $2b$

5. Solve for y: $2x + 3y = 7$
   $3x - 2y = 4$

   (A) 6
   (B) $5\frac{4}{5}$
   (C) 2
   (D) 1
   (E) $5\frac{1}{3}$

6. Solve for x: $3x + 2y = 5a + b$
   $4x - 3y = a + 7b$

   (A) $a + b$
   (B) $a - b$
   (C) $2a + b$
   (D) $17a + 17b$
   (E) $4a - 6b$

7. Solve for x: $8x^2 + 7x = 6x + 4x^2$

   (A) $-\frac{1}{4}$

   (B) 0 and $\frac{1}{4}$

   (C) 0

   (D) 0 and $-\frac{1}{4}$

   (E) none of these

8. Solve for x: $x^2 + 9x - 36 = 0$

   (A) –12 and +3
   (B) +12 and –3
   (C) –12 and –3
   (D) 12 and 3
   (E) none of these

9. Solve for x: $\sqrt{x^2 + 3} = x + 1$

   (A) ±1
   (B) 1
   (C) –1
   (D) 2
   (E) no solution

10. Solve for x: $2\sqrt{x} = -10$

    (A) 25
    (B) –25
    (C) 5
    (D) –5
    (E) no solution

SOLUTIONS TO PRACTICE EXERCISES

Diagnostic Test

1. **(D)** $(+4) + (-6) = -2$

2. **(B)** An odd number of negative signs gives a negative product.

   $(-\cancel{3})(+\cancel{4^2})\left(-\dfrac{1}{\cancel{2}}\right)\left(-\dfrac{1}{\cancel{3}}\right) = -2$

3. **(D)** The product of (-12) and $\left(+\dfrac{1}{4}\right)$ is -3.

   The product of (-18) and $\left(-\dfrac{1}{3}\right)$ is 6.

   $-\dfrac{3}{6} = -\dfrac{1}{2}$

4. **(C)** $ax + b = cx + d$

   $ax - cx = d - b$

   $(a - c)x = d - b$

   $x = \dfrac{d - b}{a - c}$

5. **(B)** Multiply the first equation by 3, the second by 7, and subtract.

   $21x - 6y = 6$
   $21x + 28y = 210$
   $-34y = -204$
   $y = 6$

6. **(C)** Add the two equations.

   $x + y = a$
   $x - y = b$
   $2x = a + b$
   $x = \dfrac{1}{2}(a + b)$

7. **(D)** $2x(2x - 1) = 0$

   $2x = 0 \qquad 2x - 1 = 0$

   $x = 0$ or $\dfrac{1}{2}$

8. **(D)** $(x - 7)(x + 3) = 0$

   $x - 7 = 0 \qquad x + 3 = 0$

   $x = 7$ or -3

9. **(E)** $\sqrt{x+1} - 3 = -7$

   $\sqrt{x+1} = -4$

   $x + 1 = 16$

   $x = 15$

   Checking, $\sqrt{16} - 3 = -7$, which is not true.

10. **(B)** $\sqrt{x^2 + 7} - 1 = x$

    $\sqrt{x^2 + 7} = x + 1$

    $x^2 + 7 = x^2 + 2x + 1$

    $7 = 2x + 1$

    $6 = 2x$

    $x = 3$

    Checking, $\sqrt{16} - 1 = 3$, which is true.

Exercise 1

1. (D) $(-4) + (+7) = +3$

2. (B) $(29{,}002) - (-1286) = 30{,}288$

3. (D) An even number of negative signs gives a positive product.

   $6 \times 4 \times 4 \times 2 = 192$

4. (B) $\dfrac{+4 + 0 + (-1) + (-5) + (-8)}{5} = \dfrac{-10}{5}$

5. (A) $5(-2) - 4(-10) - 3(5) =$
   $-10 + 40 - 15 =$
   $+15$

Exercise 2

1. (B) $3x - 2 = 3 + 2x$
   $x = 5$

2. (D) $8 - 4a + 4 = 2 + 12 - 3a$
   $12 - 4a = 14 - 3a$
   $-2 = a$

3. (A) Multiply by 8 to clear fractions.

   $y + 48 = 2y$
   $48 = y$

4. (B) Multiply by 100 to clear decimals.

   $2(x - 2) = 100$
   $2x - 4 = 100$
   $2x = 104$
   $x = 52$

5. (A) $4x + 4r = 2x + 10r$
   $2x = 14r$
   $x = 7r$

Exercise 3

1. (C) Multiply first equation by 3, then add.

$$3x - 9y = 9$$
$$\underline{2x + 9y = 11}$$
$$5x = 20$$
$$x = 4$$

2. (B) Multiply each equation by 10, then add.

$$6x + 2y = 22$$
$$\underline{5x - 2y = 11}$$
$$11x = 33$$
$$x = 3$$

3. (B) Multiply first equation by 3, second by 2, then subtract.

$$6x + 9y = 36b$$
$$\underline{6x - 2y = 14b}$$
$$11y = 22b$$
$$y = 2b$$

4. (A) $2x - 3y = 0$
$5x + y = 34$

Multiply first equation by 5, second by 2, and subtract.

$$10x - 15y = 0$$
$$\underline{10x + 2y = 68}$$
$$-17y = -68$$
$$y = 4$$

5. (B) Subtract equations.

$$x + y = -1$$
$$\underline{x - y = 3}$$
$$2y = -4$$
$$y = -2$$

Exercise 4

1. (B) $(x - 10)(x + 2) = 0$

$x - 10 = 0 \qquad x + 2 = 0$
$x = 10 \text{ or } -2$

2. (B) $(5x - 2)(5x + 2) = 0$

$5x - 2 = 0 \qquad 5x + 2 = 0$
$x = \dfrac{2}{5} \text{ or } -\dfrac{2}{5}$

3. (D) $6x(x - 7) = 0$

$6x = 0 \qquad x - 7 = 0$
$x = 0 \text{ or } 7$

4. (E) $(x - 16)(x - 3) = 0$

$x - 16 = 0 \qquad x - 3 = 0$
$x = 16 \text{ or } 3$

5. (D) $x^2 = 27$

$x = \pm\sqrt{27}$

But $\sqrt{27} = \sqrt{9} \cdot \sqrt{3} = 3\sqrt{3}$

Therefore, $x = \pm 3\sqrt{3}$

Exercise 5

1. (C) $\sqrt{2y} = 4$

    $2y = 16$

    $y = 8$

    Checking, $\sqrt{16} = 4$, which is true.

2. (D) $4\sqrt{2x-1} = 12$

    $\sqrt{2x-1} = 3$

    $2x - 1 = 9$

    $2x = 10$

    $x = 5$

    Checking, $4\sqrt{9} = 12$, which is true.

3. (E) $x^2 - 35 = 25 - 10x + x^2$

    $-35 = 25 - 10x$

    $10x = 60$

    $x = 6$

    Checking, $\sqrt{1} = 5 - 6$, which is not true.

4. (B) $18 = 3\sqrt{2y}$

    $6 = \sqrt{2y}$

    $36 = 2y$

    $y = 18$

    Checking $26 = 3\sqrt{36} + 8$,

    $26 = 3(6) + 8$, which is true.

5. (D) $\dfrac{2x}{5} = 16$

    $2x = 80$

    $x = 40$

    Checking, $\sqrt{\dfrac{80}{5}} = \sqrt{16} = 4$, which is true.

Retest

1. (D) $(-3) + (+7) - (-5) = (+9)$

2. (D) An odd number of negative signs gives a negative product.

    $\left(-\dfrac{1}{2}\right)(-4^2)(+12^2)\left(-\dfrac{1}{6}\right) = -4$

3. (D) The sum of (-4) and (-5) is (-9). The product of 9 and $-\dfrac{1}{27}$ is $-\dfrac{1}{3}$.

    $\dfrac{-9}{-\dfrac{1}{3}} = +27$

4. (B) $7b + 5d = 5x - 3b$

    $10b + 5d = 5x$

    $x = 2b + d$

5. (D) Multiply first equation by 3, second by 2, then subtract.

    $6x + 9y = 21$

    $\underline{6x - 4y = 8}$

    $13y = 13$

    $y = 1$

6. (A) Multiply first equation by 3, second by 2, then add.

    $9x + 6y = 15a + 3b$

    $\underline{8x - 6y = 2a + 14b}$

    $17x = 17a + 17b$

    $x = a + b$

7. (D) $4x^2 + x = 0$

   $x(4x + 1) = 0$

   $x = 0$ $\hspace{3cm}$ $4x + 1 = 0$

   $x = 0 \text{ or } -\dfrac{1}{4}$

8. (A) $(x + 12)(x - 3) = 0$

   $x + 12 = 0$ $\hspace{2cm}$ $x - 3 = 0$

   $x = -12 \text{ or } +3$

9. (B) $\sqrt{x^2 + 3} = x + 1$

   $\hspace{0.5cm} x^2 + 3 = x^2 + 2x + 1$

   $\hspace{1.5cm} 3 = 2x + 1$

   $\hspace{1.5cm} 2 = 2x$

   $\hspace{1.5cm} x = 1$

   Checking, $\sqrt{4} = 1 + 1$, which is true.

10. (E) $2\sqrt{x} = -10$

    $\hspace{0.5cm} \sqrt{x} = -5$

    $\hspace{0.6cm} x = 25$

    Checking, $2\sqrt{25} = -10$, which is not true.

Literal Expressions

9

DIAGNOSTIC TEST

Directions: Work out each problem. Circle the letter that appears before your answer.

Answers are at the end of the chapter.

1. If one book costs c dollars, what is the cost, in dollars, of m books?

   (A) $m + c$

   (B) $\dfrac{m}{c}$

   (C) $\dfrac{c}{m}$

   (D) mc

   (E) $\dfrac{mc}{100}$

2. Represent the cost, in dollars, of k pounds of apples at c cents per pound.

   (A) kc

   (B) $100kc$

   (C) $\dfrac{kc}{100}$

   (D) $100k + c$

   (E) $\dfrac{k}{100} + c$

3. If p pencils cost c cents, what is the cost of one pencil?

   (A) $\dfrac{c}{p}$

   (B) $\dfrac{p}{c}$

   (C) pc

   (D) $p - c$

   (E) $p + c$

4. Express the number of miles covered by a train in one hour if it covers r miles in h hours.

   (A) rh

   (B) $\dfrac{h}{r}$

   (C) $\dfrac{r}{h}$

   (D) $r + h$

   (E) $r - h$

5. Express the number of minutes in h hours and m minutes.

   (A) mh

   (B) $\dfrac{h}{60} + m$

   (C) $60(h + m)$

   (D) $\dfrac{h + m}{60}$

   (E) $60h + m$

6. Express the number of seats in the school auditorium if there are r rows with s seats each and s rows with r seats each.

   (A) $2rs$

   (B) $2r + 2s$

   (C) $rs + 2$

   (D) $2r + s$

   (E) $r + 2s$

7. How many dimes are there in n nickels and q quarters?

(A) $10nq$

(B) $\dfrac{n+q}{10}$

(C) $\dfrac{1}{2}n+\dfrac{5}{2}q$

(D) $10n + 10q$

(E) $2n+\dfrac{q}{10}$

8. Roger rents a car at a cost of D dollars per day plus c cents per mile. How many dollars must he pay if he uses the car for 5 days and drives 1000 miles?

(A) $5D + 1000c$

(B) $5D + \dfrac{c}{1000}$

(C) $5D + 100c$

(D) $5D + 10c$

(E) $5D + c$

9. The cost of a long-distance telephone call is c cents for the first three minutes and m cents for each additional minute. Represent the price of a call lasting d minutes if d is more than 3.

(A) $c + md$

(B) $c + md - 3m$

(C) $c + md + 3m$

(D) $c + 3md$

(E) cmd

10. The sales tax in Morgan County is $m\%$. Represent the total cost of an article priced at \$D.

(A) $D + mD$

(B) $D + 100mD$

(C) $D + \dfrac{mD}{100}$

(D) $D + \dfrac{m}{100}$

(E) $D + 100m$

1. COMMUNICATING WITH LETTERS

Many students who have no trouble computing with numbers panic at the sight of letters. If you understand the concepts of a problem in which numbers are given, you simply need to apply the same concepts to letters. The computational processes are exactly the same. Just figure out what you would do if you had numbers and do exactly the same thing with the given letters.

Example:

Express the number of inches in y yards, f feet, and i inches.

Solution:

We must change everything to inches and add. Since a yard contains 36 inches, y yards will contain $36y$ inches. Since a foot contains 12 inches, f feet will contain $12f$ inches. The total number of inches is $36y + 12f + i$.

Example:

Find the number of cents in $2x - 1$ dimes.

Solution:

To change dimes to cents we must multiply by 10. Think that 7 dimes would be 7 times 10 or 70 cents. Therefore the number of cents in $2x - 1$ dimes is $10(2x - 1)$ or $20x - 10$.

Example:

Find the total cost of sending a telegram of w words if the charge is c cents for the first 15 words and d cents for each additional word, if w is greater than 15.

Solution:

To the basic charge of c cents, we must add d for each word over 15. Therefore, we add d for ($w -$ 15) words. The total charge is $c + d(w - 15)$ or $c + dw - 15d$.

Example:

Kevin bought d dozen apples at c cents per apple and had 20 cents left. Represent the number of cents he had before this purchase.

Solution:

In d dozen, there are $12d$ apples. $12d$ apples at c cents each cost $12dc$ cents. Adding this to the 20 cents he has left, we find he started with $12dc + 20$ cents.

Exercise 1

Work out each problem. Circle the letter that appears before your answer.

1. Express the number of days in w weeks and w days.

   (A) $7w^2$
   (B) $8w$
   (C) $7w$
   (D) $7 + 2w$
   (E) w^2

2. The charge on the Newport Ferrry is D dollars for the car and driver and m cents for each additional passenger. Find the charge, in dollars, for a car containing four people.

   (A) $D + .03m$
   (B) $D + 3m$
   (C) $D + 4m$
   (D) $D + 300m$
   (E) $D + 400m$

3. If g gallons of gasoline cost m dollars, express the cost of r gallons.

   (A) $\dfrac{mr}{g}$
   (B) $\dfrac{rg}{m}$
   (C) rmg
   (D) $\dfrac{mg}{r}$
   (E) $\dfrac{m}{rg}$

4. How many quarters are equivalent to n nickels and d dimes?

   (A) $5n + 10d$
   (B) $25n + 50d$
   (C) $\dfrac{n+d}{25}$
   (D) $25n + 25d$
   (E) $\dfrac{n+2d}{5}$

5. A salesman earns a base salary of $100 per week plus a 5% commission on all sales over $500. Find his total earnings in a week in which he sells r dollars worth of merchandise, with r being greater than 500.

   (A) $125 + .05r$
   (B) $75 + .05r$
   (C) $125r$
   (D) $100 + .05r$
   (E) $100 - .05r$

RETEST

Work out each problem. Circle the letter that appears before your answer.

1. If a school consists of b boys, g girls, and t teachers, represent the number of students in each class if each class contains the same number of students. (Assume that there is one teacher per class.)

   (A) $\dfrac{b+g}{t}$

   (B) $t(b+g)$

   (C) $\dfrac{b}{t}+g$

   (D) $bt+g$

   (E) $\dfrac{bg}{t}$

2. Represent the total cost, in cents, of b books at D dollars each and r books at c cents each.

   (A) $\dfrac{bD}{100}+rc$

   (B) $\dfrac{bD+rc}{100}$

   (C) $100bD+rc$

   (D) $bD+100rc$

   (E) $bD+\dfrac{rc}{100}$

3. Represent the number of feet in y yards, f feet, and i inches.

   (A) $\dfrac{y}{3}+f+12i$

   (B) $\dfrac{y}{3}+f+\dfrac{i}{12}$

   (C) $3y+f+i$

   (D) $3y+f+\dfrac{i}{12}$

   (E) $3y+f+12i$

4. In a group of m men, b men earn D dollars per week and the rest earn half that amount each. Represent the total number of dollars paid to these men in a week.

   (A) $bD+b-m$

   (B) $\dfrac{1}{2}D(b+m)$

   (C) $\dfrac{3}{2}bD+mD$

   (D) $\dfrac{3}{2}D(b+m)$

   (E) $bD+\dfrac{1}{2}mD$

5. Ken bought d dozen roses for r dollars. Represent the cost of one rose.

   (A) $\dfrac{r}{d}$

   (B) $\dfrac{d}{r}$

   (C) $\dfrac{12d}{r}$

   (D) $\dfrac{12r}{d}$

   (E) $\dfrac{r}{12d}$

6. The cost of mailing a package is c cents for the first b ounces and d cents for each additional ounce. Find the cost, in cents, for mailing a package weighing f ounces if f is more than b.

   (A) $(c+d)(f-b)$

   (B) $c+d(f-b)$

   (C) $c+bd$

   (D) $c+(d-b)$

   (E) $b+(f-b)$

7. Josh's allowance is m cents per week. Represent the number of dollars he gets in a year.

   (A) $\dfrac{3m}{25}$

   (B) $5200m$

   (C) $1200m$

   (D) $\dfrac{13m}{25}$

   (E) $\dfrac{25m}{13}$

8. If it takes T tablespoons of coffee to make c cups, how many tablespoons of coffee are needed to make d cups?

   (A) $\dfrac{Tc}{d}$

   (B) $\dfrac{T}{dc}$

   (C) $\dfrac{Td}{c}$

   (D) $\dfrac{d}{Tc}$

   (E) $\dfrac{cd}{T}$

9. The charge for renting a rowboat on Loon Lake is D dollars per hour plus c cents for each minute into the next hour. How many dollars will Mr. Wilson pay if he used a boat from 3:40 P.M. to 6:20 P.M.?

   (A) $D + 40c$
   (B) $2D + 40c$
   (C) $2D + 4c$
   (D) $2D + .4c$
   (E) $D + .4c$

10. The cost for developing and printing a roll of film is c cents for processing the roll and d cents for each print. How much will it cost, in cents, to develop and print a roll of film with 20 exposures?

    (A) $20c + d$
    (B) $20(c + d)$
    (C) $c + 20d$
    (D) $c + \dfrac{d}{20}$
    (E) $\dfrac{c + d}{20}$

SOLUTIONS TO PRACTICE EXERCISES

Diagnostic Test

1. **(D)** This can be solved by a proportion, comparing books to dollars.

$$\frac{1}{c} = \frac{m}{x}$$

$$x = mc$$

2. **(C)** The cost in cents of k pounds at c cents per pound is kc. To convert this to dollars, we divide by 100.

3. **(A)** This can be solved by a proportion, comparing pencils to cents.

$$\frac{p}{c} = \frac{1}{x}$$

$$x = \frac{c}{p}$$

4. **(C)** This can be solved by a proportion, comparing miles to hours.

$$\frac{r}{h} = \frac{x}{1}$$

$$\frac{r}{h} = x$$

5. **(E)** There are 60 minutes in an hour. In h hours there are $60h$ minutes. With m additional minutes, the total is $60h + m$.

6. **(A)** r rows with s seats each have a total of rs seats. s rows with r seats each have a total of sr seats. Therefore, the school auditorium has a total of $rs + sr$ or $2rs$ seats.

7. **(C)** In n nickels, there are $5n$ cents. In q quarters, there are $25q$ cents. Altogether we have $5n + 25q$ cents. To see how many dimes this is, divide by 10.

$$\frac{5n + 25q}{10} = \frac{n + 5q}{2} = \frac{1}{2}n + \frac{5}{2}q$$

8. **(D)** The daily charge for 5 days at D dollars per day is $5D$ dollars. The charge, in cents, for 1000 miles at c cents per mile is $1000c$ cents. To change this to dollars, we divide by 100 and get $10c$ dollars. Therefore, the total cost in dollars is $5D + 10c$.

9. **(B)** The cost for the first 3 minutes is c cents. The number of additional minutes is $(d - 3)$ and the cost at m cents for each additional minute is thus $m(d - 3)$ or $md - 3m$. Therefore, the total cost is $c + md - 3m$.

10. **(C)** The sales tax is $\frac{m}{100} \cdot D$ or $\frac{mD}{100}$. Therefore, the total cost is $D + \frac{mD}{100}$.

Exercise 1

1. (B) There are 7 days in a week. *w* weeks contain 7*w* days. With *w* additional days, the total number of days is 8*w*.

2. (A) The charge is *D* dollars for car and driver. The three additional persons pay *m* cents each, for a total of 3*m* cents. To change this to dollars, divide by 100, for a total of $\frac{3m}{100}$ dollars. This can be written in decimal form as .03*m*. The total charge in dollars is then *D* + .03*m*.

3. (A) This can be solved by a proportion, comparing gallons to dollars.

   $$\frac{g}{m} = \frac{r}{x}$$
   $$gx = mr$$
   $$x = \frac{mr}{g}$$

4. (E) In *n* nickels, there are 5*n* cents. In *d* dimes, there are 10*d* cents. Altogether, we have 5*n* + 10*d* cents. To see how many quarters this gives, divide by 25.

   $\frac{5n+10d}{25} = \frac{n+2d}{5}$, since a fraction can be simplified when *every* term is divided by the same factor, in this case 5.

5. (B) Commission is paid on (*r* – 500) dollars. His commission is .05(*r* – 500) or .05*r* – 25. When this is added to his base salary of 100, we have 100 + .05*r* – 25, or 75 + .05*r*.

Retest

1. (A) The total number of boys and girls is *b* + *g*. Since there are *t* teachers, and thus *t* classes, the number of students in each class is $\frac{b+g}{t}$.

2. (C) The cost, in dollars, of *b* books at *D* dollars each is *bD* dollars. To change this to cents, we multiply by 100 and get 100*bD* cents. The cost of *r* books at *c* cents each is *rc* cents. Therefore, the total cost, in cents, is 100*bD* + *rc*.

3. (D) In *y* yards there are 3*y* feet. In *i* inches there are $\frac{i}{12}$ feet. Therefore, the total number of feet is $3y + f + \frac{i}{12}$.

4. (B) The money earned by *b* men at *D* dollars per week is *bD* dollars. The number of men remaining is (*m* – *b*), and since they earn $\frac{1}{2} D$ dollars per week, the money they earn is $\frac{1}{2} D(m-b) = \frac{1}{2} mD - \frac{1}{2} bD$. Therefore, the total amount earned is $bD + \frac{1}{2} mD - \frac{1}{2} bD = \frac{1}{2} bD + \frac{1}{2} mD = \frac{1}{2} D(b+m)$.

5. (E) This can be solved by a proportion, comparing roses to dollars. Since *d* dozen roses equals 12*d* roses,

   $$\frac{12d}{r} = \frac{1}{x}$$
   $$12d \cdot x = r$$
   $$x = \frac{r}{12d}$$

6. **(B)** The cost for the first b ounces is c cents. The number of additional ounces is $(f-b)$, and the cost at d cents for each additional ounce is $(f-b)d$. Therefore, the total cost is $c + d(f-b)$.

7. **(D)** Since there are 52 weeks in a year, his allowance in cents is $52m$. To change to dollars, we divide by 100 and get $\dfrac{52m}{100}$ or $\dfrac{13m}{25}$.

8. **(C)** This can be solved by a proportion comparing tablespoons to cups.

   $$\frac{T}{c} = \frac{x}{d}$$
   $$cx = Td$$
   $$x = \frac{Td}{c}$$

9. **(D)** The amount of time from 3:40 P.M. to 6:20 P.M. is 2 hrs. 40 min. Therefore, the charge at D dollars per hour and c cents per minute into the next hour is $2D$ dollars $+ 40c$ cents or $2D + .4c$ dollars.

10. **(C)** The cost for processing the roll is c cents. The cost for printing 20 exposures at d cents per print is $20d$ cents. Therefore, the total cost is $c + 20d$.

Roots and Radicals

10

DIAGNOSTIC TEST

Directions: Work out each problem. Circle the letter that appears before your answer.

Answers are at the end of the chapter.

1. The sum of $\sqrt{75}$ and $\sqrt{12}$ is
   - (A) $\sqrt{87}$
   - (B) $7\sqrt{3}$
   - (C) $3\sqrt{5}+3\sqrt{2}$
   - (D) $29\sqrt{3}$
   - (E) $3\sqrt{3}$

2. The difference between $\sqrt{125}$ and $\sqrt{45}$ is
   - (A) $4\sqrt{5}$
   - (B) $2\sqrt{5}$
   - (C) 2
   - (D) $5\sqrt{2}$
   - (E) 10

3. The product of $\sqrt{9x}$ and $\sqrt{4x}$ is
   - (A) $6\sqrt{x}$
   - (B) $36\sqrt{x}$
   - (C) $36x$
   - (D) $6x$
   - (E) $6x^2$

4. If $\dfrac{2}{x}=\sqrt{.16}$, then x equals
   - (A) 50
   - (B) 5
   - (C) $.5$
   - (D) $.05$
   - (E) $.005$

5. The square root of 17,956 is exactly
   - (A) 132
   - (B) 133
   - (C) 134
   - (D) 135
   - (E) 137

6. The square root of 139.24 is exactly
   - (A) 1.18
   - (B) 11.8
   - (C) 118
   - (D) $.118$
   - (E) 1180

7. Find $\sqrt{\dfrac{x^2}{36}+\dfrac{x^2}{25}}$.
   - (A) $\dfrac{11x}{30}$
   - (B) $\dfrac{9x}{30}$
   - (C) $\dfrac{x}{11}$
   - (D) $\dfrac{2x}{11}$
   - (E) $x\dfrac{\sqrt{61}}{30}$

8. $\sqrt{x^2+y^2}$ is equal to
   - (A) $x+y$
   - (B) $x-y$
   - (C) $(x+y)(x-y)$
   - (D) $\sqrt{x^2}+\sqrt{y^2}$
   - (E) none of these

9. Divide $8\sqrt{12}$ by $2\sqrt{3}$.

    (A) 16

    (B) 9

    (C) 8

    (D) 12

    (E) 96

10. $(\sqrt{2})^5$ is equal to

    (A) 2

    (B) $2\sqrt{2}$

    (C) 4

    (D) $4\sqrt{2}$

    (E) 8

1. ADDITION AND SUBTRACTION OF RADICALS

The conditions under which radicals can be added or subtracted are much the same as the conditions for letters in an algebraic expression. The radicals act as a label, or unit, and must therefore be exactly the same. In adding or subtracting, we add or subtract the coefficients, or rational parts, and carry the radical along as a label, which does not change.

Example:

$\sqrt{2} + \sqrt{3}$ cannot be added

$\sqrt{2} + \sqrt[3]{2}$ cannot be added

$4\sqrt{2} + 5\sqrt{2} = 9\sqrt{2}$

Often, when radicals to be added or subtracted are not the same, simplification of one or more radicals will make them the same. To simplify a radical, we remove any perfect square factors from underneath the radical sign.

Example:

$\sqrt{12} = \sqrt{4} \cdot \sqrt{3} = 2\sqrt{3}$

$\sqrt{27} = \sqrt{9} \cdot \sqrt{3} = 3\sqrt{3}$

If we wish to add $\sqrt{12} + \sqrt{27}$, we must first simplify each one. Adding the simplified radicals gives a sum of $5\sqrt{3}$.

Example:

$\sqrt{125} + \sqrt{20} - \sqrt{500}$

Solution:

$\sqrt{25} \cdot \sqrt{5} + \sqrt{4} \cdot \sqrt{5} - \sqrt{100} \cdot \sqrt{5}$

$= 5\sqrt{5} + 2\sqrt{5} - 10\sqrt{5}$

$= -3\sqrt{5}$

Exercise 1

Work out each problem. Circle the letter that appears before your answer.

1. Combine $4\sqrt{27} - 2\sqrt{48} + \sqrt{147}$

   (A) $27\sqrt{3}$
   (B) $-3\sqrt{3}$
   (C) $9\sqrt{3}$
   (D) $10\sqrt{3}$
   (E) $11\sqrt{3}$

2. Combine $\sqrt{80} + \sqrt{45} - \sqrt{20}$

   (A) $9\sqrt{5}$
   (B) $5\sqrt{5}$
   (C) $-\sqrt{5}$
   (D) $3\sqrt{5}$
   (E) $-2\sqrt{5}$

3. Combine $6\sqrt{5} + 3\sqrt{2} - 4\sqrt{5} + \sqrt{2}$

   (A) 8
   (B) $2\sqrt{5} + 3\sqrt{2}$
   (C) $2\sqrt{5} + 4\sqrt{2}$
   (D) $5\sqrt{7}$
   (E) 5

4. Combine $\frac{1}{2} \cdot \sqrt{180} + \frac{1}{3} \cdot \sqrt{45} - \frac{2}{5} \cdot \sqrt{20}$

   (A) $3\sqrt{10} + \sqrt{15} + 2\sqrt{2}$
   (B) $\frac{16}{5}\sqrt{5}$
   (C) $\sqrt{97}$
   (D) $\frac{24}{5}\sqrt{5}$
   (E) none of these

5. Combine $5\sqrt{mn} - 3\sqrt{mn} - 2\sqrt{mn}$

   (A) 0
   (B) 1
   (C) $\sqrt{mn}$
   (D) mn
   (E) $-\sqrt{mn}$

2. MULTIPLICATION AND DIVISION OF RADICALS

In multiplication and division, we again treat the radicals as we would treat letters in an algebraic expression. They are factors and must be treated as such.

Example:

$$\sqrt{2} \cdot \sqrt{3} = \sqrt{6}$$

Example:

$$4\sqrt{2} \cdot 5\sqrt{3} = 20 \cdot \sqrt{6}$$

Example:

$$(3\sqrt{2})^2 = 3\sqrt{2} \cdot 3\sqrt{2} = 9 \cdot 2 = 18$$

Example:

$$\frac{\sqrt{8}}{\sqrt{2}} = \sqrt{4} = 2$$

Example:

$$\frac{10\sqrt{20}}{2\sqrt{4}} = 5\sqrt{5}$$

Example:

$$\sqrt{2}(\sqrt{8} + \sqrt{18}) = \sqrt{16} + \sqrt{36} = 4 + 6 = 10$$

Exercise 2

Work out each problem. Circle the letter that appears before your answer.

1. Multiply and simplify: $2\sqrt{18} \cdot 6\sqrt{2}$

   (A) 72
   (B) 48
   (C) $12\sqrt{6}$
   (D) $8\sqrt{6}$
   (E) 36

2. Find $\left(3\sqrt{3}\right)^3$

   (A) $27\sqrt{3}$
   (B) $81\sqrt{3}$
   (C) 81
   (D) $9\sqrt{3}$
   (E) 243

3. Multiply and simplify: $\frac{1}{2}\sqrt{2}(\sqrt{6} + \frac{1}{2}\sqrt{2})$

   (A) $\sqrt{3} + \frac{1}{2}$
   (B) $\frac{1}{2} \cdot \sqrt{3}$
   (C) $\sqrt{6} + 1$
   (D) $\sqrt{6} + \frac{1}{.2}$
   (E) $\sqrt{6} + 2$

4. Divide and simplify: $\frac{\sqrt{32b^3}}{\sqrt{8b}}$

   (A) $2\sqrt{b}$
   (B) $\sqrt{2b}$
   (C) $2b$
   (D) $\sqrt{2b^2}$
   (E) $b\sqrt{2b}$

5. Divide and simplify: $\frac{15\sqrt{96}}{5\sqrt{2}}$

   (A) $7\sqrt{3}$
   (B) $7\sqrt{12}$
   (C) $11\sqrt{3}$
   (D) $12\sqrt{3}$
   (E) $40\sqrt{3}$

3. SIMPLIFYING RADICALS CONTAINING A SUM OR DIFFERENCE

In simplifying radicals that contain several terms under the radical sign, we must combine terms before taking the square root.

Example:

$$\sqrt{16+9} = \sqrt{25} = 5$$

It is not true that $\sqrt{16+9} = \sqrt{16} + \sqrt{9}$, which would be 4 + 3, or 7.

Example:

$$\sqrt{\frac{x^2}{16} - \frac{x^2}{25}} = \sqrt{\frac{25x^2 - 16x^2}{400}} = \sqrt{\frac{9x^2}{400}} = \frac{3x}{20}$$

Exercise 3

Work out each problem. Circle the letter that appears before your answer.

1. Simplify $\sqrt{\dfrac{x^2}{9} + \dfrac{x^2}{16}}$

    (A) $\dfrac{25x^2}{144}$

    (B) $\dfrac{5x}{12}$

    (C) $\dfrac{5x^2}{12}$

    (D) $\dfrac{x}{7}$

    (E) $\dfrac{7x}{12}$

2. Simplify $\sqrt{36y^2 + 64x^2}$

    (A) $6y + 8x$

    (B) $10xy$

    (C) $6y^2 + 8x^2$

    (D) $10x^2y^2$

    (E) cannot be done

3. Simplify $\sqrt{\dfrac{x^2}{64} - \dfrac{x^2}{100}}$

    (A) $\dfrac{x}{40}$

    (B) $-\dfrac{x}{2}$

    (C) $\dfrac{x}{2}$

    (D) $\dfrac{3x}{40}$

    (E) $\dfrac{3x}{80}$

4. Simplify $\sqrt{\dfrac{y^2}{2} - \dfrac{y^2}{18}}$

    (A) $\dfrac{2y}{3}$

    (B) $\dfrac{y}{5}$

    (C) $\dfrac{10y}{3}$

    (D) $\dfrac{y\sqrt{3}}{6}$

    (E) cannot be done

5. $\sqrt{a^2 + b^2}$ is equal to

    (A) $a + b$

    (B) $a - b$

    (C) $\sqrt{a^2} + \sqrt{b^2}$

    (D) $(a + b)(a - b)$

    (E) none of these

4. FINDING THE SQUARE ROOT OF A NUMBER

In finding the square root of a number, the first step is to pair off the digits in the square root sign in each direction from the decimal point. If there is an odd number of digits *before* the decimal point, insert a zero at the *beginning* of the number in order to pair digits. If there is an odd number of digits *after* the decimal point, add a zero at the *end*. It should be clearly understood that these zeros are place holders only and in no way change the value of the number. Every *pair* of numbers in the radical sign gives one digit of the square root.

Example:

Find the number of digits in the square root of 328,329.

Solution:

Pair the numbers beginning at the decimal point.

$\sqrt{32\ \ 83\ \ 29}$.

Each pair will give one digit in the square root. Therefore the square root of 328,329 has three digits.

If we were asked to find the square root of 328,329, we would look among the multiple-choice answers for a three-digit number. If there were more than one, we would have to use additional criteria for selection. Since our number ends in 9, its square root must end in a digit that, when multiplied by itself, ends in 9. Going through the digits from 0 to 9, this could be 3 ($3 \cdot 3 = 9$) or 7 ($7 \cdot 7 = 49$). Only one of these would appear among the choices, as this examination will not call for extensive computation, but rather for sound mathematical reasoning.

Example:

The square root of 4624 is exactly

(A)   64

(B)   65

(C)   66

(D)   67

(E)   68

Solution:

Since all choices contain two digits, we must reason using the last digit. It must be a number that, when multiplied by itself, will end in 4. Among the choices, the only possibility is 68 as 64^2 will end in 6, 65^2 will end in 5, 66^2 in 6, and 67^2 in 9.

Exercise 4

Work out each problem. Circle the letter that appears before your answer.

1. The square root of 17,689 is exactly

   (A) 131
   (B) 132
   (C) 133
   (D) 134
   (E) 136

2. The number of digits in the square root of 64,048,009 is

   (A) 4
   (B) 5
   (C) 6
   (D) 7
   (E) 8

3. The square root of 222.01 is exactly

   (A) 14.3
   (B) 14.4
   (C) 14.6
   (D) 14.8
   (E) 14.9

4. The square root of 25.6036 is exactly

   (A) 5.6
   (B) 5.06
   (C) 5.006
   (D) 5.0006
   (E) 5.00006

5. Which of the following square roots can be found exactly?

   (A) $\sqrt{.4}$
   (B) $\sqrt{.9}$
   (C) $\sqrt{.09}$
   (D) $\sqrt{.02}$
   (E) $\sqrt{.025}$

RETEST

Work out each problem. Circle the letter that appears before your answer.

1. The sum of $2\sqrt{8}$, $4\sqrt{50}$, and $3\sqrt{18}$ is

   (A) $33\sqrt{6}$

   (B) $9\sqrt{76}$

   (C) $33\sqrt{2}$

   (D) $135\sqrt{6}$

   (E) $136\sqrt{2}$

2. The difference between $\frac{1}{2}\sqrt{180}$ and $\frac{2}{5}\sqrt{20}$ is

   (A) $\frac{1}{10}\sqrt{160}$

   (B) $16\frac{2}{5}\sqrt{5}$

   (C) $16\frac{2}{5}$

   (D) $\frac{11}{5}\sqrt{5}$

   (E) $\frac{2}{5}\sqrt{5}$

3. The product of $a\sqrt{2x}$ and $x\sqrt{6x}$ is

   (A) $2ax^2\sqrt{3}$

   (B) $12ax^3$

   (C) $(2ax)^2\sqrt{3}$

   (D) $12ax^2$

   (E) $12ax$

4. Divide $42\sqrt{40r^3t^6}$ by $3\sqrt{5rt^2}$

   (A) $56rt^2\sqrt{2}$

   (B) $28rt\sqrt{2rt}$

   (C) $28rt^2\sqrt{2}$

   (D) $28rt\sqrt{2t}$

   (E) $56rt\sqrt{2t}$

5. Solve for x: $\frac{3}{x} = \sqrt{.09}$

   (A) 10

   (B) 1

   (C) .1

   (D) .01

   (E) 1.1

6. Find $\sqrt{\dfrac{a^2}{b^2} + \dfrac{a^2}{b^2}}$

   (A) $\frac{a^2}{b^2}$

   (B) $\frac{a}{b}$

   (C) $\frac{2a}{b}$

   (D) $\frac{a\sqrt{2}}{b}$

   (E) $\frac{a\sqrt{2}}{b^2}$

7. The square root of 213.16 is exactly

   (A) 14.2

   (B) 14.3

   (C) 14.8

   (D) 14.9

   (E) 14.6

8. The number of digits in the square root of 14,161 is

   (A) 5

   (B) 4

   (C) 3

   (D) 2

   (E) 6

9. $(2\sqrt{3})^5$ is equal to

   (A) $32\sqrt{3}$

   (B) $288\sqrt{3}$

   (C) $10\sqrt{3}$

   (D) $90\sqrt{3}$

   (E) $16\sqrt{3}$

10. Find $\sqrt{\dfrac{25m^4}{36c^{64}d^{16}}}$

    (A) $\frac{5m^2}{6c^8d^4}$

    (B) $\frac{5m^2}{6c^{32}d^4}$

    (C) $\frac{5m^2}{6c^{32}d^8}$

    (D) $\frac{5m^2}{6c^8d^8}$

    (E) $\frac{5m}{6c^{16}d^4}$

SOLUTIONS TO PRACTICE EXERCISES

Diagnostic Test

1. (B) $\sqrt{75} = \sqrt{25} \cdot \sqrt{3} = 5\sqrt{3}$

   $\sqrt{12} = \sqrt{4} \cdot \sqrt{3} = 2\sqrt{3}$

   $5\sqrt{3} + 2\sqrt{3} = 7\sqrt{3}$

2. (B) $\sqrt{125} = \sqrt{25} \cdot \sqrt{5} = 5\sqrt{5}$

   $\sqrt{45} = \sqrt{9} \cdot \sqrt{5} = 3\sqrt{5}$

   $5\sqrt{5} - 3\sqrt{5} = 2\sqrt{5}$

3. (D) $\sqrt{9x} \cdot \sqrt{4x} = \sqrt{36x^2} = 6x$

4. (B) $\sqrt{.16} = .4$

   $\dfrac{2}{x} = .4$ Multiply by x.

   $2 = .4x$

   $x = 5$

5. (C) Since the last digit is 6, the square root must end in 4 or 6.

6. (B) Since the number has two digits to the right of the decimal point, its square root will have one digit to the right of the decimal point.

7. (E) $\sqrt{\dfrac{25x^2 + 36x^2}{900}} = \sqrt{\dfrac{61x^2}{900}} = x\dfrac{\sqrt{61}}{30}$

8. (E) It is not possible to find the square root of separate terms.

9. (C) $\dfrac{8\sqrt{12}}{2\sqrt{3}} = 4\sqrt{4} = 4 \cdot 2 = 8$

10. (D) $(\sqrt{2})(\sqrt{2}) = 2$. Therefore,

    $(\sqrt{2}) \cdot (\sqrt{2}) \cdot \sqrt{2} \cdot \sqrt{2} \cdot \sqrt{2} = 4\sqrt{2}$

Exercise 1

1. (E) $4\sqrt{27} = 4\sqrt{9} \cdot \sqrt{3} = 12\sqrt{3}$

   $2\sqrt{48} = 2\sqrt{16} \cdot \sqrt{3} = 8\sqrt{3}$

   $\sqrt{147} = \sqrt{49} \cdot \sqrt{3} = 7\sqrt{3}$

   $12\sqrt{3} - 8\sqrt{3} + 7\sqrt{3} = 11\sqrt{3}$

2. (B) $\sqrt{80} = \sqrt{16} \cdot \sqrt{5} = 4\sqrt{5}$

   $\sqrt{45} = \sqrt{9} \cdot \sqrt{5} = 3\sqrt{5}$

   $\sqrt{20} = \sqrt{4} \cdot \sqrt{5} = 2\sqrt{5}$

   $4\sqrt{5} + 3\sqrt{5} - 2\sqrt{5} = 5\sqrt{5}$

3. (C) Only terms with the same radical may be combined.

   $6\sqrt{5} - 4\sqrt{5} = 2\sqrt{5}$

   $3\sqrt{2} + \sqrt{2} = 4\sqrt{2}$

   Therefore we have $2\sqrt{5} + 4\sqrt{2}$

4. (B) $\dfrac{1}{2} \cdot \sqrt{180} = \dfrac{1}{2} \cdot \sqrt{36} \cdot \sqrt{5} = 3\sqrt{5}$

   $\dfrac{1}{3} \cdot \sqrt{45} = \dfrac{1}{3} \cdot \sqrt{9} \cdot \sqrt{5} = \sqrt{5}$

   $\dfrac{2}{5} \cdot \sqrt{20} = \dfrac{2}{5} \cdot \sqrt{4} \cdot \sqrt{5} = \dfrac{4}{5}\sqrt{5}$

   $3\sqrt{5} + \sqrt{5} - \dfrac{4}{5} \cdot \sqrt{5} = 4\sqrt{5} - \dfrac{4}{5}\sqrt{5}$

   $= 3\dfrac{1}{5}\sqrt{5} = \dfrac{16}{5}\sqrt{5}$

5. (A) $5\sqrt{mn} - 5\sqrt{mn} = 0$

Exercise 2

1. **(A)** $2\sqrt{18}\cdot 6\sqrt{2}=12\sqrt{36}=12\cdot 6=72$

2. **(B)** $3\sqrt{3}\cdot 3\sqrt{3}\cdot 3\sqrt{3}=27(3\sqrt{3})=81\cdot\sqrt{3}$

3. **(A)** Using the distributive law, we have
$$\frac{1}{2}\sqrt{12}+\frac{1}{4}\cdot 2=\frac{1}{2}\sqrt{4}\cdot\sqrt{3}+\frac{1}{2}=\sqrt{3}+\frac{1}{2}$$

4. **(C)** Dividing the numbers in the radical sign, we have $\sqrt{4b^2}=2b$

5. **(D)** $3\sqrt{48}=3\sqrt{16}\cdot\sqrt{3}=12\sqrt{3}$

Exercise 3

1. **(B)** $\sqrt{\dfrac{16x^2+9x^2}{144}}=\sqrt{\dfrac{25x^2}{144}}=\dfrac{5x}{12}$

2. **(E)** The terms cannot be combined and it is not possible to take the square root of separated terms.

3. **(D)** $\sqrt{\dfrac{100x^2-64x^2}{6400}}=\sqrt{\dfrac{36x^2}{6400}}=\dfrac{6x}{80}=\dfrac{3x}{40}$

4. **(A)** $\sqrt{\dfrac{18y^2-2y^2}{36}}=\sqrt{\dfrac{16y^2}{36}}=\dfrac{4y}{6}=\dfrac{2y}{3}$

5. **(E)** It is not possible to find the square root of separate terms.

Exercise 4

1. (C) Since the last digit is 9, the square root must end in 3 or 7.

2. (A) Every pair of digits in the given number gives one digit of the square root.

3. (E) Since the number ends in 1, its square root must end in 1 or 9.

4. (B) Since the number has four digits to the right of the decimal point, its square root will have two digits to the right of the decimal point.

5. (C) In order to take the square root of a decimal, it must have an even number of decimal places so that its square root will have exactly half as many. In addition to this, the digits must form a perfect square $(\sqrt{.09} = .3)$.

Retest

1. (C) $2\sqrt{8} = 2\sqrt{4} \cdot \sqrt{2} = 4\sqrt{2}$

   $4\sqrt{50} = 4\sqrt{25} \cdot \sqrt{2} = 20\sqrt{2}$
   $3\sqrt{18} = 3\sqrt{9} \cdot \sqrt{2} = 9\sqrt{2}$
   $4\sqrt{2} + 20\sqrt{2} + 9\sqrt{2} = 33\sqrt{2}$

2. (D) $\frac{1}{2}\sqrt{180} = \frac{1}{2}\sqrt{36} \cdot \sqrt{5} = 3\sqrt{5}$

   $\frac{2}{5}\sqrt{20} = \frac{2}{5}\sqrt{4} \cdot \sqrt{5} = \frac{4}{5}\sqrt{5}$

   $3\sqrt{5} - \frac{4}{5}\sqrt{5} = \frac{11}{5}\sqrt{5}$

3. (A) $a\sqrt{2x} \cdot x\sqrt{6x} = ax\sqrt{12x^2} = 2ax^2\sqrt{3}$

4. (C) $\frac{42\sqrt{40r^3t^6}}{3\sqrt{5rt^2}} = 14\sqrt{8r^2t^4}$

   $14\sqrt{8r^2t^4} = 28rt^2\sqrt{2}$

5. (A) $\sqrt{.09} = .3$

   $\frac{3}{x} = .3$ 　　　　　　　Multiply by x.

   $3 = .3x$
   $x = 10$

6. (D) $\sqrt{\frac{2a^2}{b^2}} = \frac{a\sqrt{2}}{b}$

7. (E) Since the last digit is 6, the square root must end in 4 or 6.

8. (C) A five-digit number has a three-digit square root.

9. (B)

   $2\sqrt{3} \cdot 2\sqrt{3} \cdot 2\sqrt{3} \cdot 2\sqrt{3} \cdot 2\sqrt{3} = 32(9\sqrt{3}) = 288\sqrt{3}$

10. (C) $\sqrt{\frac{25m^4}{36c^{64}d^{16}}} = \frac{5m^2}{6c^{32}d^8}$

Factoring and Algebraic Fractions

11

DIAGNOSTIC TEST

Directions: Work out each problem. Circle the letter that appears before your answer.

Answers are at the end of the chapter.

1. Find the sum of $\frac{n}{4}$ and $\frac{2n}{3}$.

    (A) $\frac{2n^2}{7}$

    (B) $\frac{3n}{7}$

    (C) $\frac{11n}{12}$

    (D) $\frac{2n^2}{12}$

    (E) $\frac{9n}{12}$

2. Combine into a single fraction: $2 - \frac{a}{b}$.

    (A) $\frac{2-a}{b}$

    (B) $\frac{2-a}{2-b}$

    (C) $\frac{a-2b}{b}$

    (D) $\frac{2b-a}{b}$

    (E) $\frac{2a-b}{b}$

3. Divide $\frac{x-5}{x+5}$ by $\frac{5-x}{5+x}$.

    (A) 1

    (B) −1

    (C) $\frac{(x-5)^2}{(x+5)^2}$

    (D) $-\frac{(x-5)^2}{(x+5)^2}$

    (E) 0

4. Find an expression equivalent to $\left(\frac{3x^2}{y}\right)^3$.

    (A) $\frac{27x^5}{3y}$

    (B) $\frac{9x^6}{y^3}$

    (C) $\frac{9x^5}{y^3}$

    (D) $\frac{27x^5}{y^3}$

    (E) $\frac{27x^6}{y^3}$

5. Simplify $2 + \dfrac{\dfrac{1}{a}}{\dfrac{b}{a}}$

    (A) $\frac{2a+1}{b}$

    (B) $\frac{2a+1}{a}$

    (C) $\frac{2a+1}{ab}$

    (D) $\frac{4a^2+1}{xy}$

    (E) $\frac{2b+1}{b}$

155

6. Simplify $\dfrac{\dfrac{1}{a} - \dfrac{1}{b}}{2}$

  (A) $\dfrac{b-a}{2}$

  (B) $\dfrac{a-b}{2}$

  (C) $\dfrac{b-a}{2ab}$

  (D) $\dfrac{ba}{2}$

  (E) $\dfrac{2ab}{b+a}$

7. If $x + y = 16$ and $x^2 - y^2 = 48$, then $x - y$ equals

  (A) 3
  (B) 32
  (C) 4
  (D) 36
  (E) 6

8. If $(x + y)^2 = 100$ and $xy = 20$, find $x^2 + y^2$.

  (A) 100
  (B) 20
  (C) 40
  (D) 60
  (E) 80

9. If $\dfrac{1}{x} + \dfrac{1}{y} = \dfrac{1}{2}$ and $\dfrac{1}{x} - \dfrac{1}{y} = \dfrac{1}{4}$, find $\dfrac{1}{x^2} - \dfrac{1}{y^2}$.

  (A) $\dfrac{3}{4}$

  (B) $\dfrac{1}{4}$

  (C) $\dfrac{3}{16}$

  (D) $\dfrac{1}{8}$

  (E) $\dfrac{7}{8}$

10. The trinomial $x^2 - x - 20$ is exactly divisible by

  (A) $x - 4$
  (B) $x - 10$
  (C) $x + 4$
  (D) $x - 2$
  (E) $x + 5$

1. SIMPLIFYING FRACTIONS

In simplifying fractions, we must divide the numerator and denominator by the same factor. We can multiply or divide both the numerator and denominator of a fraction by the same number without changing the value of the fraction. However, if we were to add or subtract the same number in the numerator and denominator, the value of the fraction would not remain the same. When we simplify $\frac{9}{12}$ to $\frac{3}{4}$, we are really saying that $\frac{9}{12} = \frac{3 \cdot 3}{3 \cdot 4}$ and then dividing the numerator and denominator by 3. We may not say that $\frac{9}{12} = \frac{5+4}{5+7}$ and then say that $\frac{9}{12} = \frac{4}{7}$. This is a serious error in algebra as well. $\frac{9t}{12t} = \frac{3}{4}$ because we divide numerator and denominator by $3t$. However, $\frac{9+t}{12+t}$ cannot be simplified, as there is no factor that divides into the *entire* numerator as well as the *entire* denominator. *Never cancel terms!* That is, never cancel parts of numerators or denominators containing + or – signs unless they are enclosed in parentheses as parts of factors. This is one of the most frequent student errors. Be very careful to avoid it.

Example:

Simplify $\frac{4b^2 + 8b}{3b^3 + 6b^2}$

Solution:

Factoring the numerator and denominator by removing the largest common factor in both cases, we have $\frac{4b(b+2)}{3b^2(b+2)}$

The factors common to both numerator and denominator are b and $(b + 2)$. Dividing these out, we have $\frac{4}{3b}$.

Example:

Simplify $\frac{x^2 + 6x + 8}{x^2 + x - 12}$ to simplest form.

Solution:

There are no common factors here, but both numerator and denominator may be factored as trinomials. $\frac{(x+4)(x+2)}{(x+4)(x-3)}$ gives $\frac{(x+2)}{(x-3)}$ as a final answer. Remember not to cancel the x's as they are *terms* and not *factors*.

Example:

Simplify $\frac{10 - 2x}{x^2 - 4x - 5}$ to simplest form.

Solution:

The numerator contains a common factor, while the denominator must be factored as a trinomial. $2\left(\frac{5-x}{(x-5)(x+1)}\right)$

When numbers are reversed around a minus sign, they may be turned around by factoring out a (-1). $5 - x = (-1)(x - 5)$. Doing this will enable us to simplify the fraction to $\frac{-2}{x+1}$. Remember that if the terms had been reversed around a plus sign, the factors could have been divided without factoring further, as $a + b = b + a$, by the cummutative law of addition. Subtraction, however, is not commutative, necessitating the factoring of -1.

Exercise 1

Work out each problem. Circle the letter that appears before your answer.

1. Simplify to simplest form: $\dfrac{3x^3 - 3x^2y}{9x^2 - 9xy}$

    (A) $\dfrac{x}{6}$

    (B) $\dfrac{x}{3}$

    (C) $\dfrac{2x}{3}$

    (D) 1

    (E) $\dfrac{x - y}{3}$

2. Simplify to simplest form: $\dfrac{2x - 8}{12 - 3x}$

    (A) $-\dfrac{2}{3}$

    (B) $\dfrac{2}{3}$

    (C) $-\dfrac{4}{3}$

    (D) $\dfrac{4}{3}$

    (E) $-\dfrac{3}{2}$

3. Find the value of $\dfrac{3x - y}{y - 3x}$ when $x = \dfrac{2}{7}$ and

    $y = \dfrac{3}{10}$.

    (A) $\dfrac{24}{70}$

    (B) $\dfrac{11}{70}$

    (C) 0

    (D) 1

    (E) −1

4. Simplify to simplest form: $\dfrac{b^2 + b - 12}{b^2 + 2b - 15}$

    (A) $\dfrac{4}{5}$

    (B) $-\dfrac{4}{3}$

    (C) $\dfrac{b + 4}{b + 5}$

    (D) $\dfrac{b - 4}{b - 5}$

    (E) $-\dfrac{b + 4}{b + 5}$

5. Simplify to simplest form: $\dfrac{2x + 4y}{6x + 12y}$

    (A) $\dfrac{2}{3}$

    (B) $-\dfrac{2}{3}$

    (C) $-\dfrac{1}{3}$

    (D) $\dfrac{1}{3}$

    (E) 3

2. ADDITION OR SUBTRACTION OF FRACTIONS

In adding or subtracting fractions, it is necessary to have the fractions expressed in terms of the same common denominator. When adding or subtracting two fractions, use the same shortcuts used in arithmetic. Remember that $\dfrac{a}{b} + \dfrac{c}{d} = \dfrac{ad + bc}{bd}$, and that $\dfrac{a}{b} - \dfrac{c}{d} = \dfrac{ad - bc}{bd}$. All sums or differences should be simplified to simplest form.

Example:

Add $\dfrac{3}{a} + \dfrac{2}{b}$

Solution:

Add the two cross products and put the sum over the denominator product: $\dfrac{3b + 2a}{ab}$

Example:

Add $\dfrac{2a}{3} + \dfrac{4a}{5}$

Solution:

$\dfrac{10a + 12a}{15} = \dfrac{22a}{15}$

Example:

Add $\dfrac{5a}{a+b} + \dfrac{5b}{a+b}$

Solution:

Since both fractions have the same denominator, we must simply add the numerators and put the sum over the same denominator.

$\dfrac{5a + 5b}{a+b} = \dfrac{5(a+b)}{a+b} = 5$

Example:

Subtract $\dfrac{4r - s}{6} - \dfrac{2r - 7s}{6}$

Solution:

Since both fractions have the same denominator, we subtract the numerators and place the difference over the same denominator. Be very careful of the minus sign between the fractions as it will change the sign of each term in the second numerator.

$\dfrac{4r - s - (2r - 7s)}{6} = \dfrac{4r - s - 2r + 7s}{6} = \dfrac{2r + 6s}{6} = \dfrac{2(r + 3s)}{6} = \dfrac{r + 3s}{3}$

Exercise 2

Work out each problem. Circle the letter that appears before your answer.

1. Subtract $\dfrac{6x+5y}{2x}-\dfrac{4x+y}{2x}$

   (A) $1+4y$

   (B) $4y$

   (C) $1+2y$

   (D) $\dfrac{x+2y}{x}$

   (E) $\dfrac{x+3y}{x}$

2. Add $\dfrac{3c}{c+d}+\dfrac{3d}{c+d}$

   (A) $\dfrac{6cd}{c+d}$

   (B) $\dfrac{3cd}{c+d}$

   (C) $\dfrac{3}{2}$

   (D) 3

   (E) $\dfrac{9cd}{c+d}$

3. Add $\dfrac{a}{5}+\dfrac{3a}{10}$

   (A) $\dfrac{4a}{15}$

   (B) $\dfrac{a}{2}$

   (C) $\dfrac{3a^2}{50}$

   (D) $\dfrac{2a}{25}$

   (E) $\dfrac{3a^2}{15}$

4. Add $\dfrac{x+4}{6}+\dfrac{1}{2}$

   (A) $\dfrac{x+7}{6}$

   (B) $\dfrac{x+5}{8}$

   (C) $\dfrac{x+4}{12}$

   (D) $\dfrac{x+5}{12}$

   (E) $\dfrac{x+5}{6}$

5. Subtract $\dfrac{3b}{4}-\dfrac{7b}{10}$

   (A) $-\dfrac{2b}{3}$

   (B) $\dfrac{b}{5}$

   (C) $\dfrac{b}{20}$

   (D) b

   (E) $\dfrac{2b}{3}$

3. MULTIPLICATION OR DIVISION OF FRACTIONS

In multiplying or dividing fractions, we must first factor all numerators and denominators and may then divide all factors common to any numerator and any denominator. Remember always to invert the fraction following the division sign. Where exponents are involved, they are added in multiplication and subtracted in division.

Example:

Find the product of $\dfrac{x^3}{y^2}$ and $\dfrac{y^3}{x^2}$.

Solution:

Factors common to both numerator and denominator are x^2 in the first numerator and second denominator and y^2 in the first denominator and second numerator. Dividing by these common factors, we are left with $\dfrac{x}{1} \cdot \dfrac{y}{1}$. Finally, we multiply the resulting fractions, giving an answer of xy.

Example:

Divide $\dfrac{15a^2b}{2}$ by $5a^3$.

Solution:

We invert the divisor and multiply.

$\dfrac{15a^2b}{2} \cdot \dfrac{1}{5a^3}$

We can divide the first numerator and second denominator by $5a^2$, giving $\dfrac{3b}{2} \cdot \dfrac{1}{a}$ or $\dfrac{3b}{2a}$.

Exercise 3

Work out each problem. Circle the letter that appears before your answer.

1. Find the product of $\dfrac{x^2}{y^3}$ and $\dfrac{y^4}{x^5}$

   (A) $\dfrac{y^2}{x^3}$

   (B) $\dfrac{y}{x^3}$

   (C) $\dfrac{x^3}{y}$

   (D) $\dfrac{x^8}{y^7}$

   (E) $\dfrac{x}{y}$

2. Multiply c by $\dfrac{b}{c}$

   (A) $\dfrac{b}{c^2}$

   (B) $\dfrac{c^2}{b}$

   (C) b

   (D) c

   (E) bc^2

3. Divide $\dfrac{ax}{by}$ by $\dfrac{x}{y}$

   (A) $\dfrac{ax^2}{by^2}$

   (B) $\dfrac{b}{a}$

   (C) $\dfrac{a}{b}$

   (D) $\dfrac{by^2}{ax^2}$

   (E) $\dfrac{ay}{bx}$

4. Divide $4abc$ by $\dfrac{2a^2b}{3d^2}$

   (A) $\dfrac{8a^3b^2c}{3d^2}$

   (B) $\dfrac{a}{6cd^2}$

   (C) $\dfrac{2ac}{bd^2}$

   (D) $\dfrac{6cd^2}{a}$

   (E) $\dfrac{5cd^2}{a}$

5. Divide $\dfrac{3a^2c^4}{4b^2}$ by $6ac^2$

   (A) $\dfrac{ac^2}{8b^2}$

   (B) $\dfrac{ac^2}{4b^2}$

   (C) $\dfrac{4b^2}{ac^2}$

   (D) $\dfrac{8b^2}{ac^2}$

   (E) $\dfrac{ac^2}{6b^2}$

4. COMPLEX ALGEBRAIC FRACTIONS

Complex algebraic fractions are simplified by the same methods reviewed earlier for arithmetic fractions. To eliminate the fractions within the fraction, multiply *each term* of the entire complex fraction by the lowest quantity that will eliminate them all.

Example:

Simplify $\dfrac{\dfrac{3}{x}+\dfrac{2}{y}}{6}$

Solution:

We must multiply *each term* by xy, giving $\dfrac{3y+2x}{6xy}$.

No more simplification is possible beyond this. Remember *never* to cancel terms or parts of terms. We may only simplify by dividing factors.

Exercise 4

Work out each problem. Circle the letter that appears before your answer.

1. Simplify $\dfrac{\dfrac{1}{5}-\dfrac{3}{2}}{\dfrac{3}{4}}$

   (A) $\dfrac{15}{26}$

   (B) $-\dfrac{15}{26}$

   (C) 2

   (D) $\dfrac{26}{15}$

   (E) $-\dfrac{26}{15}$

2. Simplify $\dfrac{\dfrac{a}{x^2}}{\dfrac{a^2}{x}}$

   (A) $\dfrac{x}{a}$

   (B) $\dfrac{1}{a^2x}$

   (C) $\dfrac{1}{ax}$

   (D) ax

   (E) $\dfrac{a}{x}$

3. Simplify $\dfrac{\dfrac{1}{x}-\dfrac{1}{y}}{\dfrac{1}{x}+\dfrac{1}{y}}$

   (A) $\dfrac{x-y}{x+y}$

   (B) $\dfrac{x+y}{x-y}$

   (C) $\dfrac{y-x}{x+y}$

   (D) -1

   (E) $-xy$

4. Simplify $1+\dfrac{\dfrac{1}{x}}{\dfrac{1}{y}}$

   (A) $\dfrac{x+y}{x}$

   (B) $2y$

   (C) $x+1$

   (D) $\dfrac{y+1}{x}$

   (E) $\dfrac{x+1}{y}$

5. Simplify $2+\dfrac{\dfrac{1}{t}}{\dfrac{2}{t^2}}$

   (A) t^2+t

   (B) t^3

   (C) $\dfrac{2t+1}{2}$

   (D) $t+1$

   (E) $\dfrac{4+t}{2}$

5. USING FACTORING TO FIND MISSING VALUES

Certain types of problems may involve the ability to factor in order to evaluate a given expression. In particular, you should be able to factor the difference of two perfect squares. If an expression consists of two terms that are separated by a minus sign, the expression can always be factored into two binomials, with one containing the sum of the square roots and the other their difference. This can be stated by the identity $x^2 - y^2 = (x + y)(x - y)$.

Example:

If $m^2 - n^2 = 48$ and $m + n = 12$, find $m - n$.

Solution:

Since $m^2 - n^2$ is equal to $(m + n)(m - n)$, these two factors must multiply to 48. If one of them is 12, the other must be 4.

Example:

If $(a + b)^2 = 48$ and $ab = 6$, find $a^2 + b^2$.

Solution:

$(a + b)^2$ is equal to $a^2 + 2ab + b^2$. Substituting 6 for ab, we have $a^2 + 2(6) + b^2 = 48$ and $a^2 + b^2 = 36$.

Exercise 5

Work out each problem. Circle the letter that appears before your answer.

1. If $a + b = \dfrac{1}{3}$ and $a - b = \dfrac{1}{4}$, find $a^2 - b^2$.

   (A) $\dfrac{1}{12}$

   (B) $\dfrac{1}{7}$

   (C) $\dfrac{2}{7}$

   (D) $\dfrac{1}{6}$

   (E) none of these

2. If $(a - b)^2 = 40$ and $ab = 8$, find $a^2 + b^2$.

   (A) 5
   (B) 24
   (C) 48
   (D) 56
   (E) 32

3. If $a + b = 8$ and $a^2 - b^2 = 24$, then $a - b =$

   (A) 16
   (B) 4
   (C) 3
   (D) 32
   (E) 6

4. The trinomial $x^2 + 4x - 45$ is exactly divisible by

   (A) $x + 9$
   (B) $x - 9$
   (C) $x + 5$
   (D) $x + 15$
   (E) $x - 3$

5. If $\dfrac{1}{c} - \dfrac{1}{d} = 5$ and $\dfrac{1}{c} + \dfrac{1}{d} = 3$, then $\dfrac{1}{c^2} - \dfrac{1}{d^2} =$

   (A) 16
   (B) 34
   (C) 2
   (D) 15
   (E) cannot be determined

RETEST

Work out each problem. Circle the letter that appears before your answer.

1. Find the sum of $\dfrac{2n}{5}$ and $\dfrac{n}{10}$.

    (A) $\dfrac{3n}{50}$

    (B) $\dfrac{1}{2}n$

    (C) $\dfrac{2n^2}{50}$

    (D) $\dfrac{2n^2}{10}$

    (E) $\dfrac{1}{2}n$

2. Combine into a single fraction: $\dfrac{x}{y} - 3$

    (A) $\dfrac{x-3y}{y}$

    (B) $\dfrac{x-3}{y}$

    (C) $\dfrac{x-9}{3y}$

    (D) $\dfrac{x-3y}{3}$

    (E) $\dfrac{x-3y}{3y}$

3. Divide $\dfrac{x^2+2x-8}{4+x}$ by $\dfrac{2-x}{3}$.

    (A) 3

    (B) –3

    (C) $3(x-2)$

    (D) $\dfrac{3}{2-x}$

    (E) none of these

4. Find an expression equivalent to $\left(\dfrac{5a^3}{b}\right)^3$.

    (A) $\dfrac{15a^6}{b^3}$

    (B) $\dfrac{15a^9}{b^3}$

    (C) $\dfrac{125a^6}{b^3}$

    (D) $\dfrac{125a^9}{b^3}$

    (E) $\dfrac{25a^6}{b^3}$

5. Simplify $\dfrac{3-\frac{1}{x}}{\frac{y}{x}}$

    (A) $\dfrac{2}{y}$

    (B) $\dfrac{2x}{y}$

    (C) $\dfrac{3-x}{y}$

    (D) $\dfrac{3x-1}{x}$

    (E) $\dfrac{3x-1}{y}$

6. $\dfrac{3-\frac{1}{x}}{\frac{3}{x^2}}$ is equal to

    (A) $\dfrac{x^2-x}{3}$

    (B) $\dfrac{3x^2-x}{3}$

    (C) x^2-x

    (D) $\dfrac{3x-1}{3}$

    (E) $\dfrac{3-x}{3}$

7. If $a^2 - b^2 = 100$ and $a + b = 25$, then $a - b =$

    (A) 4

    (B) 75

    (C) –4

    (D) –75

    (E) 5

8. The trinomial $x^2 - 8x - 20$ is exactly divisible by

    (A) $x - 5$

    (B) $x - 4$

    (C) $x - 2$

    (D) $x - 10$

    (E) $x - 1$

9. If $\dfrac{1}{a} - \dfrac{1}{b} = 6$ and $\dfrac{1}{a} + \dfrac{1}{b} = 5$, find $\dfrac{1}{a^2} - \dfrac{1}{b^2}$.

    (A) 30

    (B) –11

    (C) 61

    (D) 11

    (E) 1

10. If $(x - y)^2 = 30$ and $xy = 17$, find $x^2 + y^2$.

    (A) –4

    (B) 4

    (C) 13

    (D) 47

    (E) 64

SOLUTIONS TO PRACTICE EXERCISES

Diagnostic Test

1. (C) $\dfrac{3n+8n}{12}=\dfrac{11n}{12}$

2. (D) $\dfrac{2}{1}-\dfrac{a}{b}=\dfrac{2b-a}{b}$

3. (B) $\dfrac{x-5}{x+5}\cdot\dfrac{5+x}{5-x}$ cancel $x+5$'s.

   $\dfrac{x-5}{5-x}=\dfrac{x-5}{-1(x-5)}=-1$

4. (E) $\dfrac{3x^2}{y}\cdot\dfrac{3x^2}{y}\cdot\dfrac{3x^2}{y}=\dfrac{27x^6}{y^3}$

5. (E) Multiply every term by a.

   $2+\dfrac{\frac{1}{a}}{\frac{b}{a}}=2+\dfrac{1}{\cancel{a}}\cdot\dfrac{\cancel{a}}{b}$

   $=2+\dfrac{1}{b}=\dfrac{2b+1}{b}$

6. (C) Multiply every term by ab.

   $\dfrac{b-a}{2ab}$

7. (A) $x^2-y^2=(x+y)(x-y)=48$

   Substituting 16 for $x+y$, we have
   $16(x-y)=48$
   $x-y=3$

8. (D) $(x+y)^2=x^2+2xy+y^2=100$

   Substituting 20 for xy, we have
   $x^2+40+y^2=100$
   $x^2+y^2=60$

9. (D) $\left(\dfrac{1}{x}+\dfrac{1}{y}\right)\left(\dfrac{1}{x}-\dfrac{1}{y}\right)=\dfrac{1}{x^2}-\dfrac{1}{y^2}$

   $\left(\dfrac{1}{2}\right)\left(\dfrac{1}{4}\right)=\dfrac{1}{x^2}-\dfrac{1}{y^2}$

   $\dfrac{1}{8}=\dfrac{1}{x^2}-\dfrac{1}{y^2}$

10. (C) $x^2-x-20=(x-5)(x+4)$

Exercise 1

1. (B) $\dfrac{3x^2(x-y)}{9x(x-y)}=\dfrac{x}{3}$

2. (A) $\dfrac{2(x-4)}{3(4-x)}=-\dfrac{2}{3}$

3. (E) $\dfrac{3x-y}{y-3x}=-1$ regardless of the values of x and y, as long as the denominator is not 0.

4. (C) $\dfrac{(b+4)(b-3)}{(b+5)(b-3)}=\dfrac{(b+4)}{(b+5)}$

5. (D) $\dfrac{2(x+2y)}{6(x+2y)}=\dfrac{2}{6}=\dfrac{1}{3}$

Exercise 2

1. (D) $\dfrac{6x+5y}{2x}-\dfrac{4x+y}{2x}$

   $=\dfrac{6x+5y-4x-y}{2x}=\dfrac{2x+4y}{2x}$

   $=\dfrac{2(x+2y)}{2x}=\dfrac{x+2y}{x}$

2. (D) $\dfrac{3c+3d}{c+d}=\dfrac{3(c+d)}{c+d}=3$

3. (B) $\dfrac{2a+3a}{10}=\dfrac{5a}{10}=\dfrac{a}{2}$

4. (A) $\dfrac{x+4+3}{6}=\dfrac{x+7}{6}$

5. (C) $\dfrac{3b(10)-4(7b)}{4(10)}=\dfrac{30b-28b}{40}$

   $=\dfrac{2b}{40}=\dfrac{b}{20}$

Exercise 3

1. (B) Divide x^2 and y^3.

   $\dfrac{1}{1}\cdot\dfrac{y}{x^3}=\dfrac{y}{x^3}$

2. (C) $c\cdot\dfrac{b}{c}=b$

3. (C) $\dfrac{ax}{by}\cdot\dfrac{y}{x}$ Divide y and x. $\dfrac{a}{b}$

4. (D) $4abc\cdot\dfrac{3d^2}{2a^2b}$ Divide 2, a, and b.

   $2c\cdot\dfrac{3d^2}{a}=\dfrac{6cd^2}{a}$

5. (A) $\dfrac{3a^2c^4}{4b^2}\cdot\dfrac{1}{6ac^2}$ Divide 3, a, and c^2.

   $\dfrac{ac^2}{4b^2}\cdot\dfrac{1}{2}=\dfrac{ac^2}{8b^2}$

Exercise 4

1. **(E)**   Multiply every term by 20.
$$\frac{4-30}{15} = \frac{-26}{15}$$

2. **(C)**   Multiply every term by x^2.
$$\frac{a}{a^2 x} = \frac{1}{ax}$$

3. **(C)**   Multiply every term by xy.
$$\frac{y-x}{y+x}$$

4. **(A)**   Multiply every term by xy.
$$\frac{x+y}{x}$$

5. **(E)**   Multiply every term by $\frac{t^2}{2}$.
$$\frac{4+t}{2}$$

Exercise 5

1. **(A)** $(a+b)(a-b) = a^2 - b^2$
$$\left(\frac{1}{3}\right)\left(\frac{1}{4}\right) = a^2 - b^2$$
$$\frac{1}{12} = a^2 - b^2$$

2. **(D)** $(a-b)^2 = a^2 - 2ab + b^2 = 40$

    Substituting 8 for ab, we have
$$a^2 - 16 + b^2 = 40$$
$$a^2 + b^2 = 56$$

3. **(C)** $(a+b)(a-b) = a^2 - b^2$
$$8(a-b) = 24$$
$$(a-b) = 3$$

4. **(A)** $x^2 + 4x - 45 = (x+9)(x-5)$

5. **(D)** $\left(\dfrac{1}{c} - \dfrac{1}{d}\right)\left(\dfrac{1}{c} + \dfrac{1}{d}\right) = \dfrac{1}{c^2} - \dfrac{1}{d^2}$
$$(5)(3) = \frac{1}{c^2} - \frac{1}{d^2}$$
$$15 = \frac{1}{c^2} - \frac{1}{d^2}$$

Retest

1. **(B)** $\dfrac{4n+n}{10}=\dfrac{5n}{10}=\dfrac{n}{2}=\dfrac{1}{2}n$

2. **(A)** $\dfrac{x}{y}-\dfrac{3}{1}=\dfrac{x-3y}{y}$

3. **(B)** $\dfrac{x^2+2x-8}{4+x}\cdot\dfrac{3}{2-x}$

   $=\dfrac{(x+4)(x-2)}{4+x}\cdot\dfrac{3}{2-x}$

   Divide $x+4$. $\dfrac{3(x-2)}{2-x}=\dfrac{3(x-2)}{-1(x-2)}=-3$

4. **(D)** $\dfrac{5a^3}{b}\cdot\dfrac{5a^3}{b}\cdot\dfrac{5a^3}{b}=\dfrac{125a^9}{b^3}$

5. **(E)** Multiply every term by x.

   $\dfrac{3x-1}{y}$

6. **(B)** Multiply every term by x^2.

   $\dfrac{3x^2-x}{3}$

7. **(A)** $a^2-b^2=(a+b)(a-b)=100$

   Substituting 25 for $a+b$, we have

   $25(a-b)=100$

   $a-b=4$

8. **(D)** $x^2-8x-20=(x-10)(x+2)$

9. **(A)** $\left(\dfrac{1}{a}-\dfrac{1}{b}\right)\left(\dfrac{1}{a}+\dfrac{1}{b}\right)=\dfrac{1}{a^2}-\dfrac{1}{b^2}$

   $(6)(5)=\dfrac{1}{a^2}-\dfrac{1}{b^2}$

   $30=\dfrac{1}{a^2}-\dfrac{1}{b^2}$

10. **(E)** $(x-y)^2=x^2-2xy+y^2=30$

    Substituting 17 for xy, we have

    $x^2-34+y^2=30$

    $x^2+y^2=64$

Problem Solving in Algebra

12

DIAGNOSTIC TEST

Directions: Work out each problem. Circle the letter that appears before your answer.

Answers are at the end of the chapter.

1. Find three consecutive odd integers such that the sum of the first two is four times the third.

    (A) 3, 5, 7
    (B) –3, –1, 1
    (C) –11, –9, –7
    (D) –7, –5, –3
    (E) 9, 11, 13

2. Find the shortest side of a triangle whose perimeter is 64, if the ratio of two of its sides is 4 : 3 and the third side is 20 less than the sum of the other two.

    (A) 6
    (B) 18
    (C) 20
    (D) 22
    (E) 24

3. A purse contains 16 coins in dimes and quarters. If the value of the coins is $2.50, how many dimes are there?

    (A) 6
    (B) 8
    (C) 9
    (D) 10
    (E) 12

4. How many quarts of water must be added to 18 quarts of a 32% alcohol solution to dilute it to a solution that is only 12% alcohol?

    (A) 10
    (B) 14
    (C) 20
    (D) 30
    (E) 34

5. Danny drove to Yosemite Park from his home at 60 miles per hour. On his trip home, his rate was 10 miles per hour less and the trip took one hour longer. How far is his home from the park?

    (A) 65 mi.
    (B) 100 mi.
    (C) 200 mi.
    (D) 280 mi.
    (E) 300 mi.

6. Two cars leave a restaurant at the same time and travel along a straight highway in opposite directions. At the end of three hours they are 300 miles apart. Find the rate of the slower car, if one car travels at a rate 20 miles per hour faster than the other.

    (A) 30
    (B) 40
    (C) 50
    (D) 55
    (E) 60

7. The numerator of a fraction is one half the denominator. If the numerator is increased by 2 and the denominator is decreased by 2, the value of the fraction is $\frac{2}{3}$. Find the numerator of the original fraction.

    (A) 4
    (B) 8
    (C) 10
    (D) 12
    (E) 20

171

8. Darren can mow the lawn in 20 minutes, while Valerie needs 30 minutes to do the same job. How many minutes will it take them to mow the lawn if they work together?

    (A) 10
    (B) 8
    (C) 16
    (D) $6\frac{1}{2}$
    (E) 12

9. Meredith is 3 times as old as Adam. Six years from now, she will be twice as old as Adam will be then. How old is Adam now?

    (A) 6
    (B) 12
    (C) 18
    (D) 20
    (E) 24

10. Mr. Barry invested some money at 5% and an amount half as great at 4%. His total annual income from both investments was $210. Find the amount invested at 4%.

    (A) $1000
    (B) $1500
    (C) $2000
    (D) $2500
    (E) $3000

In the following sections, we will review some of the major types of algebraic problems. Although not every problem you come across will fall into one of these categories, it will help you to be thoroughly familiar with these types of problems. By practicing with the problems that follow, you will learn to translate words into mathematical equations. You should then be able to handle other types of problems confidently.

In solving verbal problems, it is most important that you read carefully and know what it is that you are trying to find. Once this is done, represent your unknown algebraically. Write the equation that translates the words of the problem into the symbols of mathematics. Solve that equation by the techniques previously reviewed.

1. COIN PROBLEMS

In solving coin problems, it is best to change the value of all monies to cents before writing an equation. Thus, the number of nickels must be multiplied by 5 to give the value in cents, dimes by 10, quarters by 25, half dollars by 50, and dollars by 100.

Example:

> Sue has $1.35, consisting of nickels and dimes. If she has 9 more nickels than dimes, how many nickels does she have?

Solution:

> Let x = the number of dimes
> $x + 9$ = the number of nickels
> $10x$ = the value of dimes in cents
> $5x + 45$ = the value of nickels in cents
> 135 = the value of money she has in cents
> $10x + 5x + 45 = 135$
> $\qquad\quad 15x = 90$
> $\qquad\qquad x = 6$

She has 6 dimes and 15 nickles.

In a problem such as this, you can be sure that 6 would be among the multiple choice answers given. You must be sure to read carefully what you are asked to find and then continue until you have found the quantity sought.

Exercise 1

Work out each problem. Circle the letter that appears before your answer.

1. Marie has $2.20 in dimes and quarters. If the number of dimes is $\frac{1}{4}$ the number of quarters, how many dimes does she have?

   (A) 2
   (B) 4
   (C) 6
   (D) 8
   (E) 10

2. Lisa has 45 coins that are worth a total of $3.50. If the coins are all nickels and dimes, how many more dimes than nickels does she have?

   (A) 5
   (B) 10
   (C) 15
   (D) 20
   (E) 25

3. A postal clerk sold 40 stamps for $5.40. Some were 10-cent stamps and some were 15-cent stamps. How many 10-cent stamps were there?

   (A) 10
   (B) 12
   (C) 20
   (D) 24
   (E) 28

4. Each of the 30 students in Homeroom 704 contributed either a nickel or a quarter to the Cancer Fund. If the total amount collected was $4.70, how many students contributed a nickel?

   (A) 10
   (B) 12
   (C) 14
   (D) 16
   (E) 18

5. In a purse containing nickels and dimes, the ratio of nickels to dimes is 3 : 4. If there are 28 coins in all, what is the value of the dimes?

   (A) 60¢
   (B) $1.12
   (C) $1.60
   (D) 12¢
   (E) $1.00

2. CONSECUTIVE INTEGER PROBLEMS

Consecutive integers are one apart and can be represented algebraically as x, $x + 1$, $x + 2$, and so on. Consecutive even and odd integers are both two apart and can be represented by x, $x + 2$, $x + 4$, and so on. *Never* try to represent consecutive odd integers by x, $x + 1$, $x + 3$, etc., for if x is odd, $x + 1$ would be even.

Example:

Find three consecutive odd integers whose sum is 219.

Solution:

Represent the integers as x, $x + 2$, and $x + 4$. Write an equation stating that their sum is 219.
$$3x + 6 = 219$$
$$3x = 213$$
$$x = 71, \text{ making the integers 71, 73, and 75.}$$

Exercise 2

Work out each problem. Circle the letter that appears before your answer.

1. If $n + 1$ is the largest of four consecutive integers, represent the sum of the four integers.

    (A) $4n + 10$
    (B) $4n - 2$
    (C) $4n - 4$
    (D) $4n - 5$
    (E) $4n - 8$

2. If n is the first of two consecutive odd integers, which equation could be used to find these integers if the difference of their squares is 120?

    (A) $(n + 1)^2 - n^2 = 120$
    (B) $n^2 - (n + 1)^2 = 120$
    (C) $n^2 - (n + 2)^2 = 120$
    (D) $(n + 2)^2 - n^2 = 120$
    (E) $[(n + 2) - n]^2 = 120$

3. Find the average of four consecutive odd integers whose sum is 112.

    (A) 25
    (B) 29
    (C) 31
    (D) 28
    (E) 30

4. Find the second of three consecutive integers if the sum of the first and third is 26.

    (A) 11
    (B) 12
    (C) 13
    (D) 14
    (E) 15

5. If $2x - 3$ is an odd integer, find the next even integer.

    (A) $2x - 5$
    (B) $2x - 4$
    (C) $2x - 2$
    (D) $2x - 1$
    (E) $2x + 1$

3. AGE PROBLEMS

In solving age problems, you are usually called upon to represent a person's age at the present time, several years from now, or several years ago. A person's age x years from now is found by adding x to his present age. A person's age x years ago is found by subtracting x from his present age.

Example:

Michelle was 15 years old y years ago. Represent her age x years from now.

Solution:

Her present age is $15 + y$. In x years, her age will be her present age plus x, or $15 + y + x$.

Example:

Jody is now 20 years old and her brother, Glenn, is 14. How many years ago was Jody three times as old as Glenn was then?

Solution:

We are comparing their ages x years ago. At that time, Jody's age $(20 - x)$ was three times Glenn's age $(14 - x)$. This can be stated as the equation

$$20 - x = 3(14 - x)$$
$$20 - x = 42 - 3x$$
$$2x = 22$$
$$x = 11$$

To check, find their ages 11 years ago. Jody was 9 while Glenn was 3. Therefore, Jody was three times as old as Glenn was then.

Exercise 3

Work out each problem. Circle the letter that appears before your answer.

1. Mark is now 4 times as old as his brother Stephen. In 1 year Mark will be 3 times as old as Stephen will be then. How old was Mark two years ago?

    (A) 2
    (B) 3
    (C) 6
    (D) 8
    (E) 9

2. Mr. Burke is 24 years older than his son Jack. In 8 years, Mr. Burke will be twice as old as Jack will be then. How old is Mr. Burke now?

    (A) 16
    (B) 24
    (C) 32
    (D) 40
    (E) 48

3. Lili is 23 years old and Melanie is 15 years old. How many years ago was Lili twice as old as Melanie?

    (A) 7
    (B) 16
    (C) 9
    (D) 5
    (E) 8

4. Two years from now, Karen's age will be $2x + 1$. Represent her age two years ago.

    (A) $2x - 4$
    (B) $2x - 1$
    (C) $2x + 3$
    (D) $2x - 3$
    (E) $2x - 2$

5. Alice is now 5 years younger than her brother Robert, whose age is $4x + 3$. Represent her age 3 years from now.

    (A) $4x - 5$
    (B) $4x - 2$
    (C) $4x$
    (D) $4x + 1$
    (E) $4x - 1$

4. INVESTMENT PROBLEMS

All interest referred to is simple interest. The annual amount of interest paid on an investment is found by multiplying the amount invested, called the principal, by the percent of interest, called the rate.

$$PRINCIPAL \cdot RATE = INTEREST\ INCOME$$

Example:

Mrs. Friedman invested some money in a bank paying 4% interest annually and a second amount, $500 less than the first, in a bank paying 6% interest. If her annual income from both investments was $50, how much money did she invest at 6%?

Solution:

Represent the two investments algebraically.

x = amount invested at 4%

$x - 500$ = amount invested at 6%

$.04x$ = annual interest from 4% investment

$.06(x - 500)$ = annual interest from 6% investment

$.04x + .06(x - 500) = 50$

Multiply by 100 to remove decimals.

$$4x + 6(x - 500) = 5000$$
$$4x + 6x - 3000 = 5000$$
$$10x = 8000$$
$$x = 800$$
$$x - 500 = 300$$

She invested $300 at 6%.

Exercise 4

Work out each problem. Circle the letter that appears before your answer.

1. Barbara invested x dollars at 3% and $400 more than this amount at 5%. Represent the annual income from the 5% investment.

   (A) $.05x$
   (B) $.05(x + 400)$
   (C) $.05x + 400$
   (D) $5x + 40000$
   (E) none of these

2. Mr. Blum invested $10,000, part at 6% and the rest at 5%. If x represents the amount invested at 6%, represent the annual income from the 5% investment.

   (A) $5(x - 10,000)$
   (B) $5(10,000 - x)$
   (C) $.05(x + 10,000)$
   (D) $.05(x - 10,000)$
   (E) $.05(10,000 - x)$

3. Dr. Kramer invested $2000 in an account paying 6% interest annually. How many more dollars must she invest at 3% so that her total annual income is 4% of her entire investment?

   (A) $120
   (B) $1000
   (C) $2000
   (D) $4000
   (E) $6000

4. Marion invested $7200, part at 4% and the rest at 5%. If the annual income from both investments was the same, find her total annual income from these investments.

   (A) $160
   (B) $320
   (C) $4000
   (D) $3200
   (E) $1200

5. Mr. Maxwell inherited some money from his father. He invested $\frac{1}{2}$ of this amount at 5%, $\frac{1}{3}$ of this amount at 6%, and the rest at 3%. If the total annual income from these investments was $300, what was the amount he inherited?

   (A) $600
   (B) $60
   (C) $2000
   (D) $3000
   (E) $6000

5. FRACTION PROBLEMS

A fraction is a ratio between two numbers. If the value of a fraction is $\frac{3}{4}$, it does not mean that the numerator is 3 and the denominator 4. The numerator and denominator could be 9 and 12, respectively, or 1.5 and 2, or 45 and 60, or an infinite number of other combinations. All we know is that the ratio of numerator to denominator will be 3 : 4. Therefore, the numerator may be represented by $3x$ and the denominator by $4x$. The fraction is then represented by $\frac{3x}{4x}$.

Example:

The value of a fraction is $\frac{2}{3}$. If one is subtracted from the numerator and added to the denominator, the value of the fraction is $\frac{1}{2}$. Find the original fraction.

Solution:

Represent the original fraction as $\frac{2x}{3x}$. If one is subtracted from the numerator and added to the denominator, the new fraction is $\frac{2x-1}{3x+1}$. The value of this new fraction is $\frac{1}{2}$.

$$\frac{2x-1}{3x+1} = \frac{1}{2}$$

Cross multiply to eliminate fractions.

$$4x - 2 = 3x + 1$$
$$x = 3$$

The original fraction is $\frac{2x}{3x}$, which is $\frac{6}{9}$.

Exercise 5

Work out each problem. Circle the letter that appears before your answer.

1. A fraction is equivalent to $\frac{4}{5}$. If the numerator is increased by 4 and the denominator is increased by 10, the value of the resulting fraction is $\frac{2}{3}$. Find the numerator of the original fraction.

   (A) 4
   (B) 5
   (C) 12
   (D) 16
   (E) 20

2. What number must be added to both the numerator and the denominator of the fraction $\frac{5}{21}$ to give a fraction equal to $\frac{3}{7}$?

   (A) 3
   (B) 4
   (C) 5
   (D) 6
   (E) 7

3. The value of a certain fraction is $\frac{3}{5}$. If both the numerator and denominator are increased by 5, the new fraction is equivalent to $\frac{7}{10}$. Find the original fraction.

   (A) $\frac{3}{5}$
   (B) $\frac{6}{10}$
   (C) $\frac{9}{15}$
   (D) $\frac{12}{20}$
   (E) $\frac{15}{25}$

4. The denominator of a certain fraction is 5 more than the numerator. If 3 is added to both numerator and denominator, the value of the new fraction is $\frac{2}{3}$. Find the original fraction.

   (A) $\frac{3}{8}$
   (B) $\frac{4}{9}$
   (C) $\frac{11}{16}$
   (D) $\frac{12}{17}$
   (E) $\frac{7}{12}$

5. The denominator of a fraction is twice as large as the numerator. If 4 is added to both the numerator and denominator, the value of the fraction is $\frac{5}{8}$. Find the denominator of the original fraction.

   (A) 6
   (B) 10
   (C) 12
   (D) 14
   (E) 16

6. MIXTURE PROBLEMS

There are two kinds of mixture problems with which you should be familiar. The first is sometimes referred to as dry mixture, in which we mix dry ingredients of different values, such as nuts or coffee. Also solved by the same method are problems dealing with tickets at different prices, and similar problems. In solving this type of problem it is best to organize the data in a chart with three rows and columns, labeled as illustrated in the following example.

Example:

Mr. Sweet wishes to mix candy worth 36 cents a pound with candy worth 52 cents a pound to make 300 pounds of a mixture worth 40 cents a pound. How many pounds of the more expensive candy should he use?

Solution:

	No. of pounds ·	Price per pound =	Total value
More expensive	x	52	$52x$
Less expensive	$300 - x$	36	$36(300 - x)$
Mixture	300	40	12000

The value of the more expensive candy plus the value of the less expensive candy must be equal to the value of the mixture. Almost all mixture problems derive their equation from adding the final column in the chart.

$52x + 36(300 - x) = 12000$

Notice that all values were computed in cents to avoid decimals.

$$52x + 10,800 - 36x = 12,000$$
$$16x = 1200$$
$$x = 75$$

He should use 75 pounds of the more expensive candy.

In solving the second type of mixture problem, we are dealing with percents instead of prices and amounts of a certain ingredient instead of values. As we did with prices, we may omit the decimal point from the percents, as long as we do it in every line of the chart.

Example:

How many quarts of pure alcohol must be added to 15 quarts of a solution that is 40% alcohol to strengthen it to a solution that is 50% alcohol?

Solution:

	No. of quarts ·	Percent Alcohol =	Amount of Alcohol
Diluted	15	40	600
Pure	x	100	$100x$
Mixture	$15 + x$	50	$50(15 + x)$

Notice that the percent of alcohol in pure alcohol is 100. If we had added pure water to weaken the solution, the percent of alcohol in pure water would have been 0. Again, the equation comes from adding the final column since the amount of alcohol in the original solution plus the amount of alcohol added must equal the amount of alcohol in the new solution.

$$600 + 100x = 50(15 + x)$$
$$600 + 100x = 750 + 50x$$
$$50x = 150$$
$$x = 3$$

3 quarts of alcohol should be added.

Exercise 6

Work out each problem. Circle the letter that appears before your answer.

1. Express, in terms of x, the value, in cents, of x pounds of 40–cent cookies and $(30 - x)$ pounds of 50-cent cookies.

   (A) $150 + 10x$
   (B) $150 - 50x$
   (C) $1500 - 10x$
   (D) $1500 - 50x$
   (E) $1500 + 10x$

2. How many pounds of nuts selling for 70 cents a pound must be mixed with 30 pounds of nuts selling at 90 cents a pound to make a mixture that will sell for 85 cents a pound?

   (A) 7.5
   (B) 10
   (C) 22.5
   (D) 40
   (E) 12

3. A container holds 10 pints of a solution which is 20% acid. If 3 quarts of pure acid are added to the container, what percent of the resulting mixture is acid?

   (A) 5
   (B) 10
   (C) 20
   (D) 50
   (E) $33\frac{1}{3}$

4. A solution of 60 quarts of sugar and water is 20% sugar. How much water must be added to make a solution that is 5% sugar?

   (A) 180 qts.
   (B) 120 qts.
   (C) 100 qts.
   (D) 80 qts.
   (E) 20 qts.

5. How much water must be evaporated from 240 pounds of a solution that is 3% alcohol to strengthen it to a solution that is 5% alcohol?

   (A) 120 lbs.
   (B) 96 lbs.
   (C) 100 lbs.
   (D) 84 lbs.
   (E) 140 lbs.

7. MOTION PROBLEMS

The fundamental relationship in all motion problems is that rate times time is equal to distance.

$$\text{RATE} \cdot \text{TIME} = \text{DISTANCE}$$

The problems at the level of this examination usually deal with a relationship between distances. Most motion problems fall into one of three categories.

A. Motion in opposite directions

This can occur when objects start at the same point and move apart, or when they start at a given distance apart and move toward each other. In either case, the distance covered by the first object plus the distance covered by the second is equal to the total distance covered. This can be shown in the following diagram.

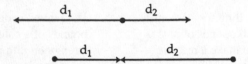

In either case, $d_1 + d_2$ = total distance covered.

B. Motion in the same direction

This type of problem is sometimes referred to as a "catch up" problem. Usually two objects leave the same place at different times and at different rates, but the one that leaves later "catches up" to the one that leaves earlier. In such cases the two distances must be equal. If one is still ahead of the other, then an equation must be written expressing this fact.

C. Round trip

In this type of problem, the rate going is different from the rate returning. The times are also different. But if we go somewhere and then return to the starting point, the distances must be equal.

To solve any type of motion problem, it is helpful to organize the information in a chart with columns for rate, time, and distance. A separate line should be used for each moving object. Be very careful of units used. If the rate is given in *miles per hour*, the time must be in *hours* and the distance will be in *miles*.

Example:

A passenger train and a freight train leave at 10:30 A.M. from stations that are 405 miles apart and travel toward each other. The rate of the passenger train is 45 miles per hour faster than that of the freight train. If they pass each other at 1:30 P.M., how fast was the passenger train traveling?

Solution:

Notice that each train traveled exactly 3 hours.

	Rate ·	Time =	Distance
Passenger	$x + 45$	3	$3x + 135$
Freight	x	3	$3x$

$$3x + 135 + 3x = 405$$
$$6x = 270$$
$$x = 45$$

The rate of the passenger train was 90 m.p.h.

Example:

Susie left her home at 11 A.M., traveling along Route 1 at 30 miles per hour. At 1 P.M., her brother Richard left home and started after her on the same road at 45 miles per hour. At what time did Richard catch up to Susie?

Solution:

	Rate	·	Time	=	Distance
Susie	30		x		$30x$
Richard	45		$x - 2$		$45x - 90$

Since Richard left 2 hours later than Susie, he traveled for $x - 2$ hours, while Susie traveled for x hours. Notice that we do not fill in 11 and 1 in the time column, as these are times on the clock and not actual hours traveled. Since Richard caught up to Susie, the distances must be equal.

$$30x = 45x - 90$$
$$90 = 15x$$
$$x = 6$$

Susie traveled for 6 hours, which means it was 6 hours past 11 A.M., or 5 P.M. when Richard caught up to her.

Example:

How far can Scott drive into the country if he drives out at 40 miles per hour and returns over the same road at 30 miles per hour and spends 8 hours away from home including a one-hour stop for lunch?

Solution:

His actual driving time is 7 hours, which must be divided into two parts. If one part is x, the other is what is left, or $7 - x$.

	Rate	·	Time	=	Distance
Going	40		x		$40x$
Return	30		$7 - x$		$210x - 30x$

The distances are equal.

$$40x = 210 - 30x$$
$$70x = 210$$
$$x = 3$$

If he traveled 40 miles per hour for 3 hours, he went 120 miles.

Exercise 7

Work out each problem. Circle the letter that appears before your answer.

1. At 10 A.M. two cars started traveling toward each other from towns 287 miles apart. They passed each other at 1:30 P.M. If the rate of the faster car exceeded the rate of the slower car by 6 miles per hour, find the rate, in miles per hour, of the faster car.

   (A) 38
   (B) 40
   (C) 44
   (D) 48
   (E) 50

2. A motorist covers 350 miles in 8 hours. Before noon he averages 50 miles per hour, but after noon he averages only 40 miles per hour. At what time did he leave?

   (A) 7 A.M.
   (B) 8 A.M.
   (C) 9 A.M.
   (D) 10 A.M.
   (E) 11 A.M.

3. At 3 P.M. a plane left Kennedy Airport for Los Angeles traveling at 600 m.p.h. At 3:30 P.M. another plane left the same airport on the same route traveling at 650 m.p.h. At what time did the second plane overtake the first?

   (A) 5:15 P.M.
   (B) 6:45 P.M.
   (C) 6:50 P.M.
   (D) 7:15 P.M.
   (E) 9:30 P.M.

4. Joe left home at 10 A.M. and walked out into the country at 4 miles per hour. He returned on the same road at 2 miles per hour. If he arrived home at 4 P.M., how many miles into the country did he walk?

   (A) 6
   (B) 8
   (C) 10
   (D) 11
   (E) 12

5. Two cars leave a restaurant at the same time and proceed in the same direction along the same route. One car averages 36 miles per hour and the other 31 miles per hour. In how many hours will the faster car be 30 miles ahead of the slower car?

   (A) 3
   (B) $3\frac{1}{2}$
   (C) 4
   (D) 6
   (E) $6\frac{1}{4}$

8. WORK PROBLEMS

In most work problems, a job is broken up into several parts, each representing a fractional portion of the entire job. For each part represented, the numerator should represent the time actually spent working, while the denominator should represent the total time needed to do the job alone. The sum of all the individual fractions must be 1 if the job is completed.

Example:

John can complete a paper route in 20 minutes. Steve can complete the same route in 30 minutes. How long will it take them to complete the route if they work together?

Solution:

	John		**Steve**		

$$\frac{\text{Time actually spent}}{\text{Time needed to do entire job alone}} \qquad \frac{x}{20} \quad + \quad \frac{x}{30} \quad = \quad 1$$

Multiply by 60 to clear fractions.

$$3x + 2x = 60$$
$$5x = 60$$
$$x = 12$$

Example:

Mr. Powell can mow his lawn twice as fast as his son Mike. Together they do the job in 20 minutes. How many minutes would it take Mr. Powell to do the job alone?

Solution:

If it takes Mr. Powell x hours to mow the lawn, Mike will take twice as long, or $2x$ hours, to mow the lawn.

$$\textbf{Mr. Powell} \qquad\qquad \textbf{Mike}$$

$$\frac{20}{x} \quad + \quad \frac{20}{2x} \quad = \quad 1$$

Multiply by $2x$ to clear fractions.

$$40 + 20 = 2x$$
$$60 = 2x$$
$$x = 30 \text{ minutes}$$

Exercise 8

Work out each problem. Circle the letter that appears before your answer.

1. Mr. White can paint his barn in 5 days. What part of the barn is still unpainted after he has worked for x days?

   (A) $\dfrac{x}{5}$

   (B) $\dfrac{5}{x}$

   (C) $\dfrac{x-5}{x}$

   (D) $\dfrac{5-x}{x}$

   (E) $\dfrac{5-x}{5}$

2. Mary can clean the house in 6 hours. Her younger sister Ruth can do the same job in 9 hours. In how many hours can they do the job if they work together?

   (A) $3\dfrac{1}{2}$

   (B) $3\dfrac{3}{5}$

   (C) 4

   (D) $4\dfrac{1}{4}$

   (E) $4\dfrac{1}{2}$

3. A swimming pool can be filled by an inlet pipe in 3 hours. It can be drained by a drainpipe in 6 hours. By mistake, both pipes are opened at the same time. If the pool is empty, in how many hours will it be filled?

   (A) 4

   (B) $4\dfrac{1}{2}$

   (C) 5

   (D) $5\dfrac{1}{2}$

   (E) 6

4. Mr. Jones can plow his field with his tractor in 4 hours. If he uses his manual plow, it takes three times as long to plow the same field. After working with the tractor for two hours, he ran out of gas and had to finish with the manual plow. How long did it take to complete the job after the tractor ran out of gas?

   (A) 4 hours

   (B) 6 hours

   (C) 7 hours

   (D) 8 hours

   (E) $8\dfrac{1}{2}$ hours

5. Michael and Barry can complete a job in 2 hours when working together. If Michael requires 6 hours to do the job alone, how many hours does Barry need to do the job alone?

   (A) 2

   (B) $2\dfrac{1}{2}$

   (C) 3

   (D) $3\dfrac{1}{2}$

   (E) 4

RETEST

Work out each problem. Circle the letter that appears before your answer.

1. Three times the first of three consecutive odd integers is 10 more than the third. Find the middle integer.

   (A) 7
   (B) 9
   (C) 11
   (D) 13
   (E) 15

2. The denominator of a fraction is three times the numerator. If 8 is added to the numerator and 6 is subtracted from the denominator, the resulting fraction is equivalent to $\frac{8}{9}$. Find the original fraction.

   (A) $\frac{16}{18}$
   (B) $\frac{1}{3}$
   (C) $\frac{8}{24}$
   (D) $\frac{5}{3}$
   (E) $\frac{8}{16}$

3. How many quarts of water must be added to 40 quarts of a 5% acid solution to dilute it to a 2% solution?

   (A) 80
   (B) 40
   (C) 60
   (D) 20
   (E) 50

4. Miriam is 11 years older than Charles. In three years she will be twice as old as Charles will be then. How old was Miriam 2 years ago?

   (A) 6
   (B) 8
   (C) 9
   (D) 17
   (E) 19

5. One printing press can print the school newspaper in 12 hours, while another press can print it in 18 hours. How long will the job take if both presses work simultaneously?

   (A) 7 hrs. 12 min.
   (B) 6 hrs. 36 min.
   (C) 6 hrs. 50 min.
   (D) 7 hrs. 20 min.
   (E) 7 hrs. 15 min.

6. Janet has $2.05 in dimes and quarters. If she has four fewer dimes than quarters, how much money does she have in dimes?

   (A) 30¢
   (B) 80¢
   (C) $1.20
   (D) 70¢
   (E) 90¢

7. Mr. Cooper invested a sum of money at 6%. He invested a second sum, $150 more than the first, at 3%. If his total annual income was $54, how much did he invest at 3%?

   (A) $700
   (B) $650
   (C) $500
   (D) $550
   (E) $600

8. Two buses are 515 miles apart. At 9:30 A.M. they start traveling toward each other at rates of 48 and 55 miles per hour. At what time will they pass each other?

   (A) 1:30 P.M.
   (B) 2:30 P.M.
   (C) 2 P.M.
   (D) 3 P.M.
   (E) 3:30 P.M.

9. Carol started from home on a trip averaging 30 miles per hour. How fast must her mother drive to catch up to her in 3 hours if she leaves 30 minutes after Carol?

   (A) 35 m.p.h.
   (B) 39 m.p.h.
   (C) 40 m.p.h.
   (D) 55 m.p.h.
   (E) 60 m.p.h.

10. Dan has twice as many pennies as Frank. If Frank wins 12 pennies from Dan, both boys will have the same number of pennies. How many pennies did Dan have originally?

   (A) 24
   (B) 12
   (C) 36
   (D) 48
   (E) 52

SOLUTIONS TO PRACTICE EXERCISES

Diagnostic Test

1. **(D)** Represent the integers as x, $x + 2$, and $x + 4$.

$$x + x + 2 = 4(x + 4)$$
$$2x + 2 = 4x + 16$$
$$-14 = 2x$$
$$x = -7, x + 2 = -5, x + 4 = -3$$

2. **(B)** Represent the first two sides as $4x$ and $3x$, then the third side is $7x - 20$.

$$4x + 3x + (7x - 20) = 64$$
$$14x - 20 = 64$$
$$14x = 84$$
$$x = 6$$

The shortest side is $3(6) = 18$.

3. **(D)**
$$\text{Let } x = \text{the number of dimes}$$
$$16 - x = \text{the number of quarters}$$
$$10x = \text{value of dimes in cents}$$
$$400 - 25x = \text{value of quarters in cents}$$
$$10x + 400 - 25x = 250$$
$$-15x = -150$$
$$x = 10$$

4. **(D)**

	No of Quarts ·	Percent Alcohol =	Amount of Alcohol
Original	18	32	576
Added	x	0	0
New	$18 + x$	12	$216 + 12x$

$$576 = 216 + 12x$$
$$360 = 12x$$
$$x = 30$$

5. **(E)**

	R ·	T =	D
Going	60	x	$60x$
Return	50	$x + 1$	$50x + 50$

$$60x = 50x + 50$$
$$10x = 50$$
$$x = 5$$

If he drove for 5 hours at 60 miles per hour, he drove 300 miles.

6. **(B)**

	R ·	T =	D
Slow	x	3	$3x$
Fast	$x + 20$	3	$3x + 60$

$$3x + 3x + 60 = 300$$
$$6x = 240$$
$$x = 40$$

7. **(C)** Represent the original fraction by $\dfrac{x}{2x}$.
$$\frac{x + 2}{2x - 2} = \frac{2}{3}$$
Cross multiply.
$$3x + 6 = 4x - 4$$
$$x = 10$$

8. **(E)**

Darren		Valerie	
$\dfrac{x}{20}$	$+$	$\dfrac{x}{30}$	$= 1$

Multiply by 60.
$$3x + 2x = 60$$
$$5x = 60$$
$$x = 12$$

9. **(A)**
$$\text{Let } x = \text{Adam's age now}$$
$$3x = \text{Meredith's age now}$$
$$x + 6 = \text{Adam's age in 6 years}$$
$$3x + 6 = \text{Meredith's age in 6 years}$$
$$3x + 6 = 2(x + 6)$$
$$3x + 6 = 2x + 12$$
$$x = 6$$

10. **(B)**
$$\text{Let } x = \text{amount invested at 4\%}$$
$$2x = \text{amount invested at 5\%}$$
$$.04x + .05(2x) = 210$$

Multiply by 100 to eliminate decimals.
$$4x + 5(2x) = 21{,}000$$
$$14x = 21{,}000$$
$$x = \$1500$$

Exercise 1

1. **(A)** Let x = number of dimes

    $4x$ = number of quarters

    $10x$ = value of dimes in cents

    $100x$ = value of quarters in cents

    $$10x + 100x = 220$$
    $$100x = 220$$
    $$x = 2$$

2. **(A)** Let x = number of nickels

    $45 - x$ = number of dimes

    $5x$ = value of nickels in cents

    $450 - 10x$ = value of dimes in cents

    $$5x + 450 - 10x = 350$$
    $$-5x = -100$$
    $$x = 20$$

    20 nickels and 25 dimes

3. **(B)** Let x = number of 10-cent stamps

    $40 - x$ = number of 15-cent stamps

    $10x$ = value of 10-cent stamps

    $600 - 15x$ = value of 15-cent stamps

    $$10x + 600 - 15x = 540$$
    $$-5x = -60$$
    $$x = 12$$

4. **(C)** Let x = number of nickels

    $30 - x$ = number of quarters

    $5x$ = value of nickels in cents

    $750 - 25x$ = value of quarters in cents

    $$5x + 750 - 25x = 470$$
    $$-20x = -280$$
    $$x = 14$$

5. **(C)** Let $3x$ = number of nickels

    $4x$ = number of dimes

    $$3x + 4x = 28$$
    $$7x = 28$$
    $$x = 4$$

    There are 16 dimes, worth $1.60.

Exercise 2

1. **(B)** Consecutive integers are 1 apart. If the fourth is $n + 1$, the third is n, the second is $n - 1$, and the first is $n - 2$. The sum of these is $4n - 2$.

2. **(D)** The other integer is $n + 2$. If a difference is positive, the larger quantity must come first.

3. **(D)** To find the average of any 4 numbers, divide their sum by 4.

4. **(C)** Represent the integers as x, $x + 1$, and $x + 2$.

    $$x + x + 2 = 26$$
    $$2x = 24$$
    $$x = 12$$
    $$x + 1 = 13$$

5. **(C)** An even integer follows an odd integer, so simply add 1.

Exercise 3

1. (C)   Let x = Stephen's age now
    $4x$ = Mark's age now
    $x+1$ = Stephen's age in 1 year
    $4x+1$ = Mark's age in 1 year
    $4x+1 = 3(x+1)$
    $4x+1 = 3x+3$
    $x = 2$

    Mark is now 8, so 2 years ago he was 6.

2. (D)   Let x = Jack's age now
    $x+24$ = Mr. Burke's age now
    $x+8$ = Jack's age in 8 years
    $x+32$ = Mr. Burke's age in 8 years
    $x+32 = 2(x+8)$
    $x+32 = 2x+16$
    $16 = x$

    Jack is now 16, Mr. Burke is 40.

3. (A)   The fastest reasoning here is from the answers. Subtract each number from both ages, to see which results in Lili being twice as old as Melanie. 7 years ago, Lili was 16 and Melanie was 8.

    Let x = number of years ago

    Then $23 - x = 2(15 - x)$
    $23 - x = 30 - 2x$
    $7 = x$

4. (D)   Karen's age now can be found by subtracting 2 from her age 2 years from now. Her present age is $2x - 1$. To find her age 2 years ago, subtract another 2.

5. (D)   Alice's present age is $4x - 2$. In 3 years her age will be $4x + 1$.

Exercise 4

1. (B)   She invested $x + 400$ dollars at 5%. The income is $.05(x + 400)$.

2. (E)   He invested $10,000 - x$ dollars at 5%. The income is $.05(10,000 - x)$.

3. (D)   Let x = amount invested at 3%
    $2000 + x$ = her total investment
    $.06(2000) + .03x = .04(2000 + x)$

    Multiply by 100 to eliminate decimals.

    $6(2000) + 3x = 4(2000 + x)$
    $12,000 + 3x = 8000 + 4x$
    $4000 = x$

4. (B)   Let x = amount invested at 4%
    $7200 - x$ = amount invested at 5%
    $.04x = .05(7200 - x)$

    Multiply by 100 to eliminate decimals.

    $4x = 5(7200 - x)$
    $4x = 36,000 - 5x$
    $9x = 36,000$
    $x = 4000$

    Her income is .04(4000) + .05(3200). This is $160 + $160, or $320.

5. (E)   In order to avoid fractions, represent his inheritance as $6x$. Then $\frac{1}{2}$ his inheritance is $3x$ and $\frac{1}{3}$ his inheritance is $2x$.

    Let $3x$ = amount invested at 5%
    $2x$ = amount invested at 6%
    x = amount invested at 3%

    $.05(3x) + .06(2x) + .03(x) = 300$

    Multiply by 100 to eliminate decimals.

    $5(3x) + 6(2x) + 3(x) = 30,000$
    $15x + 12x + 3x = 30,000$
    $30x = 30,000$
    $x = 1000$

    His inheritance was $6x$, or $6000.

Exercise 5

1. **(D)** Represent the original fraction as $\dfrac{4x}{5x}$.

   $$\dfrac{4x+4}{5x+10}=\dfrac{2}{3}$$

   Cross multiply.

   $$12x+12=10x+20$$
   $$2x=8$$
   $$x=4$$

   The original numerator was $4x$, or 16.

2. **(E)** While this can be solved using the equation $\dfrac{5+x}{21+x}=\dfrac{3}{7}$, it is probably easier to work from the answers. Try adding each choice to the numerator and denominator of $\dfrac{5}{21}$ to see which gives a result equal to $\dfrac{3}{7}$.

   $$\dfrac{5+7}{21+7}=\dfrac{12}{28}=\dfrac{3}{7}$$

3. **(C)** Here again, it is fastest to reason from the answers. Add 5 to each numerator and denominator to see which will result in a new fraction equal to $\dfrac{7}{10}$.

   $$\dfrac{9+5}{15+5}=\dfrac{14}{20}=\dfrac{7}{10}$$

4. **(E)** Here again, add 3 to each numerator and denominator of the given choices to see which will result in a new fraction equal to $\dfrac{2}{3}$.

   $$\dfrac{7+3}{12+3}=\dfrac{10}{15}=\dfrac{2}{3}$$

5. **(C)** Represent the original fraction by $\dfrac{x}{2x}$.

   $$\dfrac{x+4}{2x+4}=\dfrac{5}{8}$$

   Cross multiply.

   $$8x+32=10x+20$$
   $$12=2x$$
   $$x=6$$

   The original denominator is $2x$, or 12.

Exercise 6

1. **(C)** Multiply the number of pounds by the price per pound to get the total value.

   $$40(x)+50(30-x)=$$
   $$40x+1500-50x=$$
   $$1500-10x$$

2. **(B)**

No. of Pounds ·	Price per Pound =	Total Value
x	70	$70x$
30	90	2700
$x+30$	85	$85(x+30)$

   $$70x+2700=85(x+30)$$
   $$70x+2700=85x+2550$$
   $$150=15x$$
   $$x=10$$

3. **(D)**

	No. of Pints	% of Acid =	Amount of Acid
Original	10	.20	2
Added	6	1.00	6
New	16		8

   Remember that 3 quarts of acid are 6 pints. There are now 8 pints of acid in 16 pints of solution. Therefore, the new solution is $\dfrac{1}{2}$ or 50% acid.

4. **(A)**

No. of Quarts ·	% of Sugar =	Amount of Sugar
60	20	1200
x	0	0
$60+x$	5	$5(60+x)$

   $$1200=5(60+x)$$
   $$1200=300+5x$$
   $$900=5x$$
   $$x=180$$

5. **(B)**

No. of Pounds ·	% of Alcohol =	Amount of Sugar
240	3	720
x	0	0
$240-x$	5	$5(240-x)$

   Notice that when x quarts were evaporated, x was *subtracted* from 240 to represent the number of pounds in the mixture.

   $$720=5(240-x)$$
   $$720=1200-5x$$
   $$5x=480$$
   $$x=96$$

Exercise 7

1. (C)

	R	·	T	=	D
Slow	x		3.5		$3.5x$
Fast	$x + 6$		3.5		$3.5(x + 6)$

The cars each traveled from 10 A.M. to 1:30 P.M., which is $3\frac{1}{2}$ hours.

$3.5x + 3.5(x + 6) = 287$

Multiply by 10 to eliminate decimals.

$35x + 35(x + 6) = 2870$
$35x + 35x + 210 = 2870$
$70x = 2660$
$x = 38$

The rate of the faster car was $x + 6$ or 44 m.p.h.

2. (C)

	R	·	T	=	D
Before noon	50		x		$50x$
After noon	40		$8 - x$		$40(8 - x)$

The 8 hours must be divided into 2 parts.

$50x + 40(8 - x) = 350$
$50x + 320 - 40x = 350$
$10x = 30$
$x = 3$

If he traveled 3 hours before noon, he left at 9 A.M.

3. (E)

R	·	T	=	D
600		x		$600x$
650		$x - \frac{1}{2}$		$650(x - \frac{1}{2})$

The later plane traveled $\frac{1}{2}$ hour less.

$600x = 650\left(x - \frac{1}{2}\right)$
$600x = 650x - 325$
$325 = 50x$
$6\frac{1}{2} = x$

The plane that left at 3 P.M. traveled for $6\frac{1}{2}$ hours. The time is then 9:30 P.M.

4. (B)

	R	·	T	=	D
Going	4		x		$4x$
Return	2		$6 - x$		$2(6 - x)$

He was gone for 6 hours.

$4x = 2(6 - x)$
$4x = 12 - 2x$
$6x = 12$
$x = 2$

If he walked for 2 hours at 4 miles per hour, he walked for 8 miles.

5. (D)

R	·	T	=	D
36		x		$36x$
31		x		$31x$

They travel the same number of hours.

$36x - 31x = 30$
$5x = 30$
$x = 6$

This problem may be reasoned without an equation. If the faster car gains 5 miles per hour on the slower car, it will gain 30 miles in 6 hours.

Exercise 8

1. **(E)** In x days, he has painted $\frac{x}{5}$ of the barn. To find what part is still unpainted, subtract the part completed from 1. Think of 1 as $\frac{5}{5}$.

$$\frac{5}{5} - \frac{x}{5} = \frac{5-x}{5}$$

2. **(B)**

Mary		Ruth	
$\frac{x}{6}$	$+$	$\frac{x}{9}$	$= 1$

Multiply by 18.

$$3x + 2x = 18$$
$$5x = 18$$
$$x = 3\frac{3}{5}$$

3. **(E)**

Inlet		Drain	
$\frac{x}{3}$	$-$	$\frac{x}{6}$	$= 1$

Multiply by 6.

$$2x - x = 6$$
$$x = 6$$

Notice the two fractions are subtracted, as the drainpipe does not help the inlet pipe but works against it.

4. **(B)**

Tractor		Plow	
$\frac{2}{4}$	$+$	$\frac{x}{12}$	$= 1$

This can be done without algebra, as half the job was completed by the tractor; therefore, the second fraction must also be equal to $\frac{1}{2}$. x is therefore 6.

5. **(C)**

Michael		Barry	
$\frac{2}{6}$	$+$	$\frac{2}{x}$	$= 1$

Multiply by $6x$.

$$2x + 12 = 6x$$
$$12 = 4x$$
$$x = 3$$

Retest

1. **(B)** Represent the integers as x, $x + 2$, and $x + 4$.

$$3x = (x+4) + 10$$
$$2x = 14$$
$$x = 7$$
$$x + 2 = 9$$

2. **(C)** Represent the original fraction by $\frac{x}{3x}$.

$$\frac{x+8}{3x-6} = \frac{8}{9}$$

Cross multiply.

$$9x + 72 = 24x - 48$$
$$120 = 15x$$
$$x = 8$$
$$3x = 24$$

The original fraction is $\frac{8}{24}$.

3. **(C)**

	No. of Quarts	· Percent Alcohol =	Amount of Alcohol
Original	40	5	200
Added	x	0	0
New	$40 + x$	2	$80 + 2x$

$$200 = 80 + 2x$$
$$120 = 2x$$
$$x = 60$$

4. **(D)** Let x = Charles' age now

$$x + 11 = \text{Miriam's age now}$$
$$x + 3 = \text{Charles' age in 3 years}$$
$$x + 14 = \text{Miriam's age in 3 years}$$

$$x + 14 = 2(x + 3)$$
$$x + 14 = 2x + 6$$
$$x = 8$$

Therefore, Miriam is 19 now and 2 years ago was 17.

5. (A) Fast Press   Slow Press

$$\frac{x}{12} \quad + \quad \frac{x}{18} \quad = 1$$

Multiply by 36.

$$3x + 2x = 36$$
$$5x = 36$$
$$x = 7\frac{1}{5} \text{ hours}$$
$$= 7 \text{ hours } 12 \text{ minutes}$$

6. (A) Let x = the number of dimes
$x + 4$ = the number of quarters
$10x$ = the value of dimes in cents
$25x + 100$ = the value of quarters in cents

$$10x + 25x + 100 = 205$$
$$35x = 105$$
$$x = 3$$

She has 30¢ in dimes.

7. (A) Let x = amount invested at 6%
$x + 150$ = amount invested at 3%
$.06x + .03(x + 150) = 54$

Multiply by 100 to eliminate decimals

$$6x + 3(x + 150) = 5400$$
$$6x + 3x + 450 = 5400$$
$$9x = 4950$$
$$x = \$550$$
$$x + 150 = \$700$$

8. (B)

	R	·	T	=	D
Slow	48		x		$48x$
Fast	55		x		$55x$

$$48x + 55x = 515$$
$$103x = 515$$
$$x = 5 \text{ hours}$$

Therefore, they will pass each other 5 hours after 9:30 A.M., 2:30 P.M.

9. (A)

	R	·	T	=	D
Carol	30		3.5		105
Mother	x		3		$3x$

$$3x = 105$$
$$x = 35 \text{ m.p.h.}$$

10. (D) Let x = number of pennies Frank has
$2x$ = number of pennies Dan has

$$x + 12 = 2x - 12$$
$$x = 24$$

Therefore, Dan originally had 48 pennies.

Geometry

13

DIAGNOSTIC TEST

Directions: Work out each problem. Circle the letter that appears before your answer.

Answers are at the end of the chapter.

1. If the angles of a triangle are in the ratio 5 : 6 : 7, the triangle is

   (A) acute
   (B) isosceles
   (C) obtuse
   (D) right
   (E) equilateral

2. A circle whose area is 4 has a radius of x. Find the area of a circle whose radius is $3x$.

   (A) 12
   (B) 36
   (C) $4\sqrt{3}$
   (D) 48
   (E) 144

3. A spotlight is attached to the ceiling 2 feet from one wall of a room and 3 feet from the wall adjacent. How many feet is it from the intersection of the two walls?

   (A) 4
   (B) 5
   (C) $3\sqrt{2}$
   (D) $\sqrt{13}$
   (E) $2\sqrt{3}$

4. In parallelogram $ABCD$, angle B is 5 times as large as angle C. What is the measure in degrees of angle B?

   (A) 30
   (B) 60
   (C) 100
   (D) 120
   (E) 150

5. A rectangular box with a square base contains 24 cubic feet. If the height of the box is 18 inches, how many feet are there in each side of the base?

   (A) 1
   (B) 2
   (C) $\dfrac{2\sqrt{3}}{3}$
   (D) $\dfrac{\sqrt{3}}{2}$
   (E) $\sqrt{3}$

6. In triangle ABC, $AB = BC$. If angle B contains x degrees, find the number of degrees in angle A.

   (A) x
   (B) $180 - x$
   (C) $180 - \dfrac{x}{2}$
   (D) $90 - \dfrac{x}{2}$
   (E) $90 - x$

197

7. In the diagram below, *AB* is perpendicular to *BC*. If angle *XBY* is a straight angle and angle *XBC* contains 37°, find the number of degrees in angle *ABY*.

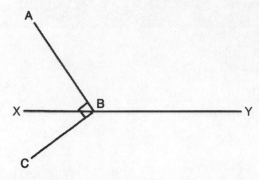

(A) 37
(B) 53
(C) 63
(D) 127
(E) 143

8. If $\overline{AB}$ is parallel to $\overline{CD}$, angle 1 contains 40°, and angle 2 contains 30°, find the number of degrees in angle *FEG*.

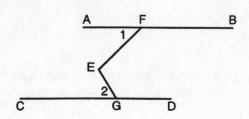

(A) 110
(B) 140
(C) 70
(D) 40
(E) 30

9. In a circle whose center is *O*, arc *AB* contains 100°. Find the number of degrees in angle *ABO*.

(A) 50
(B) 100
(C) 40
(D) 65
(E) 60

10. Find the length of the line segment joining the points whose coordinates are (–3, 1) and (5, –5).

(A) 10
(B) $2\sqrt{5}$
(C) $2\sqrt{10}$
(D) 100
(E) $\sqrt{10}$

The questions in the following area will expect you to recall some of the numerical relationships learned in geometry. If you are thoroughly familiar with these relationships, you should not find these questions difficult. As mentioned earlier, be particularly careful with units. For example, you cannot multiply a dimension given in feet by another given in inches when you are finding area. Read each question very carefully for the units given. In the following sections, all the needed formulas with illustrations and practice exercises are to help you prepare for the geometry questions on your test.

1. AREAS

A. Rectangle = base · altitude = *bh*

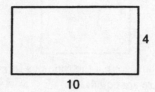

Area = 40

B. Parallelogram = base · altitude = *bh*

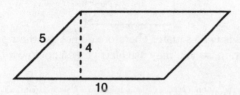

Area = 40

Notice that the altitude is different from the side. It is always shorter than the second side of the parallelogram, as a perpendicular is the shortest distance from a point to a line.

C. Rhombus = $\frac{1}{2}$ · product of the diagonals = $\frac{1}{2}d_1d_2$

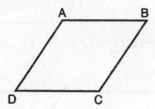

If $AC = 20$ and $BD = 30$, the area of $ABCD = \frac{1}{2}(20)(30) = 300$

D. Square = side · side = s^2

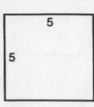

Area = 25

Remember that every square is a rhombus, so that the rhombus formula may be used for a square if the diagonal is given. The diagonals of a square are equal.

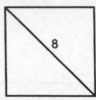

Area = $\frac{1}{2}(8)(8) = 32$

Remember also that a rhombus is *not* a square. Therefore do not use the s^2 formula for a rhombus. A rhombus, however, is a parallelogram, so you may use bh if you do not know the diagonals.

E. Triangle = $\frac{1}{2}$ · base · altitude = $\frac{1}{2}bh$

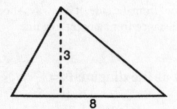

$A = \frac{1}{2}(8)(3) = 12$

F. Equilateral Triangle = $\frac{1}{4}$ · side squared · $\sqrt{3} = \frac{s^2}{4}\sqrt{3}$

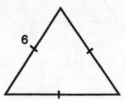

$A = \frac{36}{4}\sqrt{3} = 9\sqrt{3}$

G. Trapezoid = $\frac{1}{2}$ · altitude · sum of bases = $\frac{1}{2}h(b_1 + b_2)$

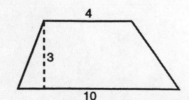

$A = \frac{1}{2}(3)(14) = 21$

H. Circle = π · radius squared = π · r^2

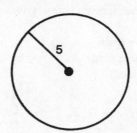

$A = \pi \cdot (5)^2 = 25\pi$

Remember that π is the ratio of the circumference of any circle and its diameter. $\pi = \frac{c}{d}$. The approximations you have used for π in the past (3.14 or $\frac{22}{7}$) are just that—approximations. π is an irrational number and cannot be expressed as a fraction or terminating decimal. Therefore all answers involving π should be left in terms of π unless you are given a specific value to substitute for π.

A word about units—Area is measured in square units. That is, we wish to compute how many squares one inch on each side (a square inch) or one foot on each side (a square foot), etc., can be used to cover a given surface. To change from square inches to square feet or square yards, remember that

144 square inches = 1 square foot

9 square feet = 1 square yard

1 square foot   **1 square yard**

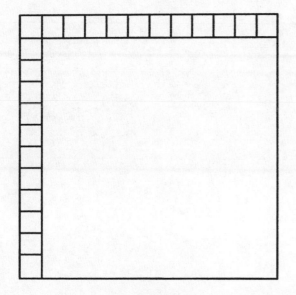

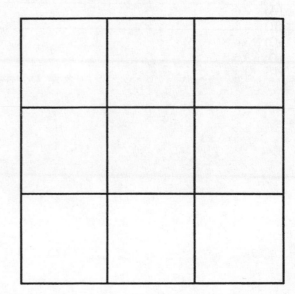

12" = 1'
12 one inch squares in a row
12 rows
144 square inches in 1 sq. ft.

3' = 1 yd.
3 one foot squares in a row
3 rows
9 square feet in 1 sq. yd.

Exercise 1

Work out each problem. Circle the letter that appears before your answer.

1. The dimensions of a rectangular living room are 18 feet by 20 feet. How many square yards of carpeting are needed to cover the floor?

    (A) 360
    (B) 42
    (C) 40
    (D) 240
    (E) 90

2. In a parallelogram whose area is 15, the base is represented by $x + 7$ and the altitude is $x - 7$. Find the base of the parallelogram.

    (A) 8
    (B) 15
    (C) 1
    (D) 34
    (E) 5

3. The sides of a right triangle are 6, 8, and 10. Find the altitude drawn to the hypotenuse.

    (A) 2.4
    (B) 4.8
    (C) 3.4
    (D) 3.5
    (E) 4.2

4. If the diagonals of a rhombus are represented by $4x$ and $6x$, the area may be represented by

    (A) $6x$
    (B) $24x$
    (C) $12x$
    (D) $6x^2$
    (E) $12x^2$

5. A circle is inscribed in a square whose side is 6. Express the area of the circle in terms of π.

    (A) 6π
    (B) 3π
    (C) 9π
    (D) 36π
    (E) 12π

2. PERIMETER

The perimeter of a figure is the distance around the outside. If you were fencing in an area, the number of feet of fencing you would need is the perimeter. Perimeter is measured in linear units, that is, centimeters, inches, feet, meters, yards, etc.

A. Any polygon = sum of all sides

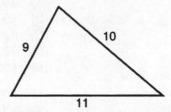

$P = 9 + 10 + 11 = 30$

B. Circle = $\pi \cdot$ diameter = πd

or

$2 \cdot \pi \cdot$ radius = $2\pi r$

Since $2r = d$, these formulas are the same. The perimeter of a circle is called its circumference.

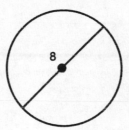

$C = \pi \cdot 8 = 8\pi$

or

$C = 2 \cdot \pi \cdot 4 = 8\pi$

The distance covered by a wheel in one revolution is equal to the circumference of the wheel. In making one revolution, every point on the rim comes in contact with the ground. The distance covered is then the same as stretching the rim out into a straight line.

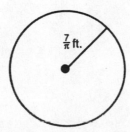

The distance covered by this wheel in one revolution is $2 \cdot \pi \cdot \dfrac{7}{\pi} = 14$ feet.

Exercise 2

Work out each problem. Circle the letter that appear before your answer.

1. The area of an equilateral triangle is $16\sqrt{3}$. Find its perimeter.

   (A) 24
   (B) 16
   (C) 48
   (D) $24\sqrt{3}$
   (E) $48\sqrt{3}$

2. The hour hand of a clock is 3 feet long. How many feet does the tip of this hand move between 9:30 P.M. and 1:30 A.M. the following day?

   (A) π
   (B) 2π
   (C) 3π
   (D) 4π
   (E) 24π

3. If the radius of a circle is increased by 3, the circumference is increased by

   (A) 3
   (B) 3π
   (C) 6
   (D) 6π
   (E) 4.5

4. The radius of a wheel is 18 inches. Find the number of feet covered by this wheel in 20 revolutions.

   (A) 360π
   (B) 360
   (C) 720π
   (D) 720
   (E) 60π

5. A square is equal in area to a rectangle whose base is 9 and whose altitude is 4. Find perimeter of the square.

   (A) 36
   (B) 26
   (C) 13
   (D) 24
   (E) none of these

3. RIGHT TRIANGLES

A. Pythagorean theorem

$(\text{leg})^2 + (\text{leg})^2 = (\text{hypotenuse})^2$

$$\left(5\right)^2 + \left(2\right)^2 = x^2$$
$$25 + 4 = x^2$$
$$29 = x^2$$
$$x = \sqrt{29}$$

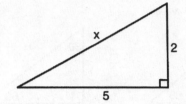

B. Pythagorean triples

These are sets of numbers that satisfy the Pythagorean Theorem. When a given set of numbers such as 3, 4, 5 forms a Pythagorean triple ($3^2 + 4^2 = 5^2$), any multiples of this set such as 6, 8, 10 or 30, 40, 50 also form a Pythagorean triple. Memorizing the sets of Pythagorean triples that follow will save you valuable time in solving problems, for, if you recognize given numbers as multiples of Pythagorean triples, you do not have to do any arithmetic at all. The most common Pythagorean triples that should be memorized are

3, 4, 5

5, 12, 13

8, 15, 17

7, 24, 25

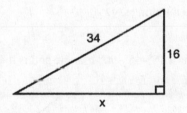

Squaring 34 and 16 to apply the Pythagorean theorem would take too much time. Instead, recognize the hypotenuse as 2(17). Suspect an 8, 15, 17 triangle. Since the given leg is 2(8), the missing leg will be 2(15) or 30, without any computation at all.

C. 30°–60°–90° triangle

a) The leg opposite the 30° angle is one-half the hypotenuse.

b) The leg opposite the 60° angle is one-half the hypotenuse $\cdot \sqrt{3}$.

c) An altitude in an equilateral triangle forms a 30°–60°–90° triangle and is therefore equal to one-half the side $\cdot \sqrt{3}$.

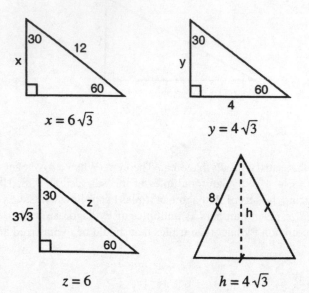

D. 45°–45°–90° triangle (isosceles right triangle)

a)   Each leg is one-half the hypotenuse times $\sqrt{2}$.

b)   Hypotenuse is leg times $\sqrt{2}$.

c)   The diagonal of a square forms a 45°–45°–90° triangle and is therefore equal to a side times $\sqrt{2}$.

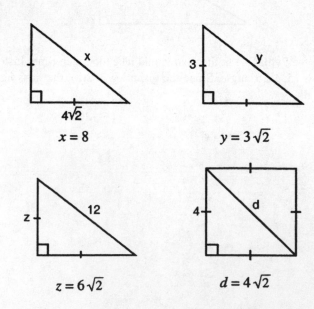

Exercise 3

Work out each problem. Circle the letter that appears before your answer.

1. A farmer uses 140 feet of fencing to enclose a rectangular field. If the ratio of length to width is 3 : 4, find the diagonal, in feet, of the field.

    (A) 50
    (B) 100
    (C) 20
    (D) 10
    (E) cannot be determined

2. Find the altitude of an equilateral triangle whose side is 20.

    (A) 10
    (B) $20\sqrt{3}$
    (C) $10\sqrt{3}$
    (D) $20\sqrt{2}$
    (E) $10\sqrt{2}$

3. Two boats leave the same dock at the same time, one traveling due west at 8 miles per hour and the other due north at 15 miles per hour. How many miles apart are the boats after three hours?

    (A) 17
    (B) 69
    (C) 75
    (D) 51
    (E) 39

4. Find the perimeter of a square whose diagonal is $6\sqrt{2}$.

    (A) 24
    (B) $12\sqrt{2}$
    (C) 12
    (D) 20
    (E) $24\sqrt{2}$

5. Find the length of *DB*.

    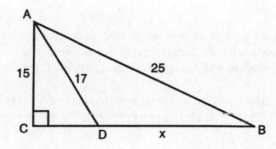

    (A) 8
    (B) 10
    (C) 12
    (D) 15
    (E) 20

4. COORDINATE GEOMETRY

A. Distance between two points =

$$\sqrt{(x_2 - x_1)^2 + (y_2 - y_1)^2}$$

The distance between (–3, 2) and (5, –1) is

$$\sqrt{[-3-5]^2 + [2-(-1)]^2} = \sqrt{(-8)^2 + (3)^2} = \sqrt{64+9} = \sqrt{73}$$

B. The midpoint of a line segment =

$$\left(\frac{x_1 + x_2}{2}, \frac{y_1 + y_2}{2} \right)$$

Since a midpoint is in the middle, its coordinates are found by averaging the x coordinates and averaging the y coordinates. Remember that to find the average of two numbers, you add them and divide by two. Be very careful of signs in adding signed numbers. Review the rules given earlier if necessary.

The midpoint of the segment joining (–4, 1) to (–2, –9) is

$$\left(\frac{-4+(-2)}{2}, \frac{1+(-9)}{2} \right) = \left(\frac{-6}{2}, \frac{-8}{2} \right) = (-3, -4)$$

Exercise 4

Work out each problem. Circle the letter that appears before your answer.

1. *AB* is the diameter of a circle whose center is *O*. If the coordinates of *A* are (2, 6) and the coordinates of *B* are (6, 2), find the coordinates of *O*.
   (A) (4, 4)
   (B) (4, –4)
   (C) (2, –2)
   (D) (0, 0)
   (E) (2, 2)

2. *AB* is the diameter of a circle whose center is *O*. If the coordinates of *O* are (2, 1) and the coordinates of *B* are (4, 6), find the coordinates of *A*.

   (A) $\left(3, 3\frac{1}{2} \right)$

   (B) $\left(1, 2\frac{1}{2} \right)$

   (C) (0, –4)

   (D) $\left(2\frac{1}{2}, 1 \right)$

   (E) $\left(-1, -2\frac{1}{2} \right)$

3. Find the distance from the point whose coordinates are (4, 3) to the point whose coordinates are (8, 6).
   (A) 5
   (B) 25
   (C) $\sqrt{7}$
   (D) $\sqrt{67}$
   (E) 15

4. The vertices of a triangle are (2, 1), (2, 5), and (5, 1). The area of the triangle is
   (A) 12
   (B) 10
   (C) 8
   (D) 6
   (E) 5

5. The area of a circle whose center is at (0,0) is 16π. The circle passes through each of the following points *except*
   (A) (4, 4)
   (B) (0, 4)
   (C) (4, 0)
   (D) (–4, 0)
   (E) (0, –4)

5. PARALLEL LINES

A. If two lines are parallel and cut by a transversal, the alternate interior angles are congruent.

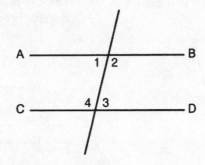

If $\overline{AB}$ is parallel to $\overline{CD}$, then   angle 1 ≅ angle 3 and
angle 2 ≅ angle 4.

B. If two parallel lines are cut by a transversal, the corresponding angles are congruent.

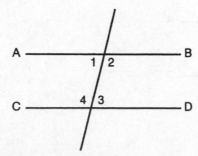

If $\overline{AB}$ is parallel to $\overline{CD}$, then   angle 1 ≅ angle 5
angle 2 ≅ angle 6
angle 3 ≅ angle 7
angle 4 ≅ angle 8

C. If two parallel lines are cut by a transversal, interior angles on the same side of the transversal are supplementary.

If $\overline{AB}$ is parallel to $\overline{CD}$,   angle 1 + angle 4 = 180°
angle 2 + angle 3 = 180°

Exercise 5

Work out each problem. Circle the letter that appears before your answer.

1. If $\overline{AB}$ is parallel to $\overline{CD}$, $\overline{BC}$ is parallel to $\overline{ED}$, and angle $B = 30°$, find the number of degrees in angle D.

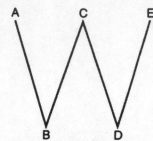

(A) 30
(B) 60
(C) 150
(D) 120
(E) none of these

2. If $\overline{AB}$ is parallel to $\overline{CD}$, angle $A = 35°$, and angle $C = 45°$, find the number of degrees in angle AEC.

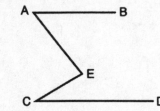

(A) 35
(B) 45
(C) 70
(D) 80
(E) 100

3. If $\overline{AB}$ is parallel to $\overline{CD}$ and angle $1 = 130°$, find angle 2.

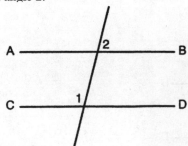

(A) 130°
(B) 100°
(C) 40°
(D) 60°
(E) 50°

4. If $\overline{AB}$ is parallel to $\overline{CD}$, $\overline{EF}$ bisects angle BEG, and $\overline{GF}$ bisects angle EGD, find the number of degrees in angle EFG.

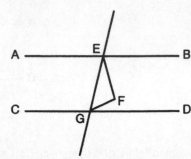

(A) 40
(B) 60
(C) 90
(D) 120
(E) cannot be determined

5. If $\overline{AB}$ is parallel to $\overline{CD}$ and angle $1 = x°$, then the sum of angle 1 and angle 2 is

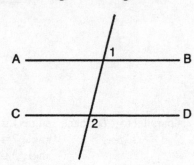

(A) $2x°$
(B) $(180 - x)°$
(C) $180°$
(D) $(180 + x)°$
(E) none of these

6. TRIANGLES

A. If two sides of a triangle are congruent, the angles opposite these sides are congruent.

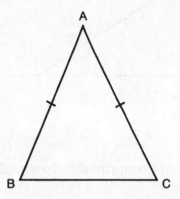

If $\overline{AB} \cong \overline{AC}$, then angle $B \cong$ angle C.

B. If two angles of a triangle are congruent, the sides opposite these angles are congruent.

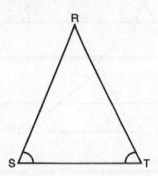

If angle $S \cong$ angle T, then $\overline{RS} \cong \overline{RT}$.

C. The sum of the measures of the angles of a triangle is 180°.

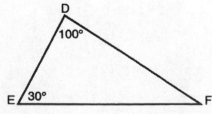

Angle $F = 180° - 100° - 30° = 50°$.

D. The measure of an exterior angle of a triangle is equal to the sum of the measures of the two remote interior angles.

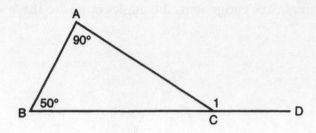

Angle 1 = 140°

E. If two angles of one triangle are congruent to two angles of a second triangle, the third angles are congruent.

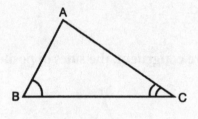

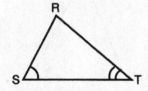

Angle *A* will be congruent to angle *R*.

Exercise 6

Work out each problem. Circle the letter that appears before your answer.

1. The angles of a triangle are in the ratio 1 : 5 : 6. This triangle is
   (A) acute
   (B) obtuse
   (C) isosceles
   (D) right
   (E) equilateral

2. If the vertex angle of an isosceles triangle is 50°, find the number of degrees in one of the base angles.
   (A) 50
   (B) 130
   (C) 60
   (D) 65
   (E) 55

3. In triangle ABC, angle A is three times as large as angle B. The exterior angle at C is 100°. Find the number of degrees in angle A.
   (A) 60
   (B) 80
   (C) 20
   (D) 25
   (E) 75

4. If a base angle of an isosceles triangle is represented by $x°$, represent the number of degrees in the vertex angle.
   (A) $180 - x$
   (B) $x - 180$
   (C) $2x - 180$
   (D) $180 - 2x$
   (E) $90 - 2x$

5. In triangle ABC, $\overline{AB} = \overline{BC}$. If angle $A = (4x - 30)°$ and angle $C = (2x + 10)°$, find the number of degrees in angle B.
   (A) 20
   (B) 40
   (C) 50
   (D) 100
   (E) 80

7. POLYGONS

A. The sum of the measures of the angles of a polygon of *n* sides is (*n* – 2)180°.

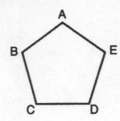

Since *ABCDE* has 5 sides, angle *A* + angle *B* + angle *C* + angle *D* + angle *E* = (5 – 2)180° = 3(180)° = 540°

B. Properties of a parallelogram

a) Opposite sides are parallel

b) Opposite sides are congruent

c) Opposite angles are congruent

d) Consecutive angles are supplementary

e) Diagonals bisect each other

C. Properties of a rectangle

a) All 5 properties of a parallelogram

b) All angles are right angles

c) Diagonals are congruent

D. Properties of a rhombus

a) All 5 properties of a parallelogram

b) All sides are congruent

c) Diagonals are perpendicular to each other

d) Diagonals bisect the angles

E. Properties of a square

a) All 5 parallelogram properties

b) Two additional rectangle properties

c) Three additional rhombus properties

Exercise 7

Work out each problem. Circle the letter that appears before your answer.

1. Find the number of degrees in the sum of the interior angles of a hexagon.

   (A) 360
   (B) 540
   (C) 720
   (D) 900
   (E) 1080

2. In parallelogram $ABCD$, $AB = x + 4$, $BC = x - 6$, and $CD = 2x - 16$. Find AD.

   (A) 20
   (B) 24
   (C) 28
   (D) 14
   (E) 10

3. In parallelogram $ABCD$, $AB = x + 8$, $BC = 3x$, and $CD = 4x - 4$. $ABCD$ must be a

   (A) rectangle
   (B) rhombus
   (C) trapezoid
   (D) square
   (E) pentagon

4. The sum of the angles in a rhombus is

   (A) 180°
   (B) 360°
   (C) 540°
   (D) 720°
   (E) 450°

5. Which of the following statements is *false?*

   (A) A square is a rhombus.
   (B) A rhombus is a parallelogram.
   (C) A rectangle is a rhombus.
   (D) A rectangle is a parallelogram.
   (E) A square is a rectangle.

8. CIRCLES

A. A central angle is equal in degrees to its intercepted arc.

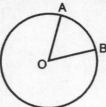

If arc $AB = 50°$, then angle $AOB = 50°$.

B. An inscribed angle is equal in degrees to one-half its intercepted arc.

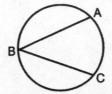

If arc $AC = 100°$, then angle $ABC = 50°$.

C. An angle formed by two chords intersecting in a circle is equal in degrees to one-half the sum of its intercepted arcs.

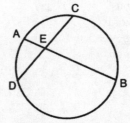

If arc $AD = 30°$ and arc $CB = 120°$, then angle $AED = 75°$.

D. An angle outside the circle formed by two secants, a secant and a tangent, or two tangents is equal in degrees to one-half the difference of its intercepted arcs.

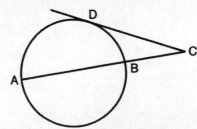

If arc $AD = 120°$ and arc $BD = 30°$, then angle $C = 45°$.

E. Two tangent segments drawn to a circle from the same external point are congruent.

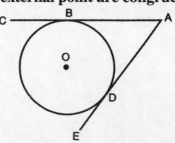

If $\overline{AC}$ and $\overline{AE}$ are tangent to circle O at B and D, then $AB \cong AD$.

Exercise 8

Work out each problem. Circle the letter that appears before your answer.

1. If circle O is inscribed in triangle ABC, find the length of side AB.

    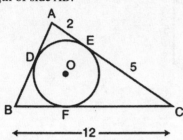

    (A) 12
    (B) 14
    (C) 9
    (D) 10
    (E) 7

2. Find angle x.

    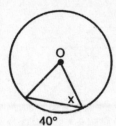

    (A) 40°
    (B) 20°
    (C) 50°
    (D) 70°
    (E) 80°

3. Find angle x.

    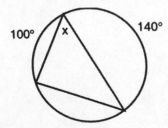

    (A) 120°
    (B) 50°
    (C) 70°
    (D) 40°
    (E) 60°

4. Find the number of degrees in arc AC.

    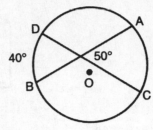

    (A) 60
    (B) 50
    (C) 25
    (D) 100
    (E) 20

5. The number of degrees in angle ABC is

    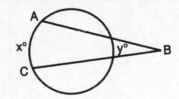

    (A) $\frac{1}{2}y$

    (B) y

    (C) $\frac{1}{2}x$

    (D) $\frac{1}{2}(x-y)$

    (E) $\frac{1}{2}(x+y)$

9. VOLUMES

A. The volume of a rectangular solid is equal to the product of its length, width, and height.

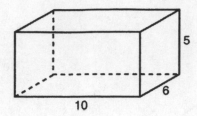

$V = (10)(6)(5) = 300$

B. The volume of a cube is equal to the cube of an edge, since the length, width, and height are all equal.

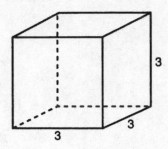

$V = (3)^3 = 27$

C. The volume of a cylinder is equal to π times the square of the radius of the base times the height.

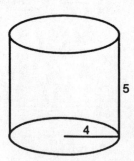

$V = \pi (4)^2 (5) = 80\pi$

Exercise 9

Work out each problem. Circle the letter that appears before your answer.

1. The surface area of a cube is 96 square feet. How many cubic feet are there in the volume of the cube?

   (A) 16
   (B) 4
   (C) 12
   (D) 64
   (E) 32

2. A cylindrical pail has a radius of 7 inches and a height of 10 inches. Approximately how many gallons will the pail hold if there are 231 cubic inches to a gallon? (Use $\pi = \frac{22}{7}$)

   (A) .9
   (B) 4.2
   (C) 6.7
   (D) 5.1
   (E) 4.8

3. Water is poured into a cylindrical tank at the rate of 9 cubic inches a minute. How many minutes will it take to fill the tank if its radius is 3 inches and its height is 14 inches? (Use $\pi = \frac{22}{7}$)

   (A) $14\frac{2}{3}$
   (B) 44
   (C) 30
   (D) $27\frac{2}{9}$
   (E) 35

4. A rectangular tank 10 inches by 8 inches by 4 inches is filled with water. If the water is to be transferred to smaller tanks in the form of cubes 4 inches on a side, how many of these tanks are needed?

   (A) 4
   (B) 5
   (C) 6
   (D) 7
   (E) 8

5. The base of a rectangular tank is 6 feet by 5 feet and its height is 16 inches. Find the number of cubic feet of water in the tank when it is $\frac{5}{8}$ full.

   (A) 25
   (B) 40
   (C) 480
   (D) 768
   (E) 300

10. SIMILAR POLYGONS

A. Corresponding angles of similar polygons are congruent.

B. Corresponding sides of similar polygons are in proportion.

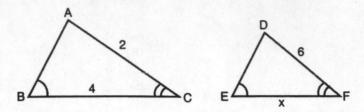

If triangle *ABC* is similar to triangle *DEF* and the sides and angles are given as marked, then *EF* must be equal to 12 as the ratio of corresponding sides is 2 : 6 or 1 : 3.

C. When figures are similar, all ratios between corresponding lines are equal. This includes the ratios of corresponding sides, medians, altitudes, angle bisectors, radii, diameters, perimeters, and circumferences. The ratio is referred to as the linear ratio or ratio of similitude.

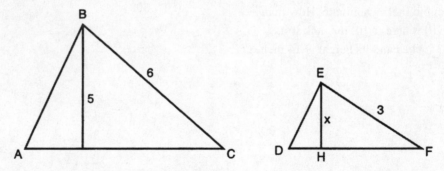

If triangle *ABC* is similar to triangle *DEF* and the segments are given as marked, then *EH* is equal to 2.5 because the linear ratio is 6 : 3 or 2 : 1.

D. When figures are similar, the ratio of their areas is equal to the square of the linear ratio.

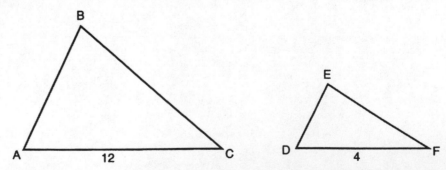

If triangle *ABC* is similar to triangle *DEF*, the area of triangle *ABC* will be 9 times as great as the area of triangle *DEF*. The linear ratio is 12 : 4 or 3 : 1. The area ratio will be the square of this or 9 : 1. If the area of triangle *ABC* had been given as 27, the area of triangle *DEF* would be 3.

E. When figures are similar, the ratio of their volumes is equal to the cube of their linear ratio.

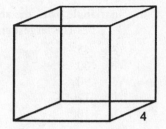 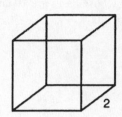

The volume of the larger cube is 8 times the volume of the smaller cube. The ratio of sides is 4 : 2 or 2 : 1. The ratio of areas would be 4 : 1. The ratio of volumes would be 8 : 1.

Exercise 10

Work out each problem. Circle the letter that appears before your answer.

1. If the area of a circle of radius x is 5π, find the area of a circle of radius $3x$.

   (A) 10π
   (B) 15π
   (C) 20π
   (D) 30π
   (E) 45π

2. If the length and width of a rectangle are each doubled, the area is increased by

   (A) 50%
   (B) 100%
   (C) 200%
   (D) 300%
   (E) 400%

3. The area of one circle is 9 times as great as the area of another. If the radius of the smaller circle is 3, find the radius of the larger circle.

   (A) 9
   (B) 12
   (C) 18
   (D) 24
   (E) 27

4. If the radius of a circle is doubled, then

   (A) the circumference and area are both doubled
   (B) the circumference is doubled and the area is multiplied by 4
   (C) the circumference is multiplied by 4 and the area is doubled
   (D) the circumference and area are each multiplied by 4
   (E) the circumference stays the same and the area is doubled

5. The volumes of two similar solids are 250 and 128. If a dimension of the larger solid is 25, find the corresponding side of the smaller solid.

   (A) 12.8
   (B) 15
   (C) 20
   (D) 40
   (E) cannot be determined

RETEST

Work out each problem. Circle the letter that appears before your answer.

1. The area of a trapezoid whose bases are 10 and 12 and whose altitude is 3 is

    (A) 66
    (B) 11
    (C) 33
    (D) 25
    (E) $16\frac{1}{2}$

2. The circumference of a circle whose area is 16π is

    (A) 8π
    (B) 4π
    (C) 16π
    (D) 8
    (E) 16

3. Find the perimeter of a square whose diagonal is 8.

    (A) 32
    (B) 16
    (C) $32\sqrt{2}$
    (D) $16\sqrt{2}$
    (E) $32\sqrt{3}$

4. The length of the line segment joining the point $A(4, -3)$ to $B(7, -7)$ is

    (A) $\sqrt{221}$
    (B) $\sqrt{185}$
    (C) 7
    (D) $6\frac{1}{2}$
    (E) 5

5. Find angle x if $\overline{AB}$ is parallel to $\overline{CD}$.

    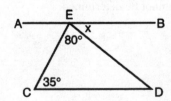

    (A) 35°
    (B) 80°
    (C) 245°
    (D) 65°
    (E) 55°

6. In triangle ABC, the angles are in a ratio of $1 : 1 : 2$. The largest angle of the triangle is

    (A) 45°
    (B) 60°
    (C) 90°
    (D) 120°
    (E) 100°

7. Find the number of degrees in each angle of a regular pentagon.

    (A) 72
    (B) 108
    (C) 60
    (D) 180
    (E) 120

8. Find the number of degrees in arc AB.

    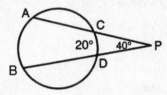

    (A) 80
    (B) 20
    (C) 60
    (D) 100
    (E) 90

9. Find the edge, in inches, of a cube whose volume is equal to the volume of a rectangular solid 2 in. by 6 in. by 18 in.

    (A) 4
    (B) 8
    (C) 5
    (D) 6
    (E) 7

10. If the volume of one cube is 8 times as great as another, then the ratio of the area of a face of the larger cube to the area of a face of the smaller cube is

    (A) $2 : 1$
    (B) $4 : 1$
    (C) $\sqrt{2} : 1$
    (D) $8 : 1$
    (E) $2\sqrt{2} : 1$

SOLUTIONS TO PRACTICE EXERCISES

Diagnostic Test

1. **(A)** Represent the angles as $5x$, $6x$, and $7x$. They must add up to $180°$.

$$18x = 180$$
$$x = 10$$

The angles are $50°$, $60°$, and $70°$, an acute triangle.

2. **(B)** The area of a circle is πr^2. The area of a circle with radius x is πx^2, which equals 4. The area of a circle with radius $3x$ is $\pi (3x)^2 = 9\pi x^2 = 9 \cdot 4 = 36$.

3. **(D)**

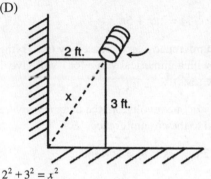

$$2^2 + 3^2 = x^2$$
$$4 + 9 = x^2$$
$$13 = x^2$$
$$\sqrt{13} = x$$

4. **(E)**

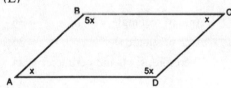

The sum of the angles in a parallelogram is $360°$.

$$12x = 360°$$
$$x = 30°$$

Angle $B = 5x = 5 \cdot 30° = 150°$

5. **(A)** The volume of a rectangular box is the product of its length, width, and height. Since the height is 18 inches, or $1\frac{1}{2}$ feet, and the length and width of the square base are the same, we have

$$x \cdot x \cdot 1\frac{1}{2} = 24$$
$$x^2 = 16$$
$$x = 4$$

6. **(D)** The remaining degrees of the triangle are $180 - x$. Since the triangle is isosceles, the remaining angles are equal, each $\dfrac{180 - x}{2} = 90 - \dfrac{x}{2}$.

7. **(D)**

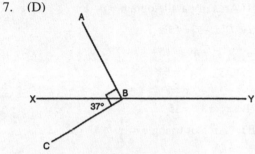

Angle $ABX = 90° - 37° = 53°$

Angle $ABY = 180° - 53° = 127°$

8. **(C)**

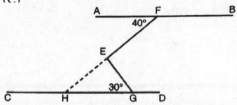

Extend $\overline{FE}$ to H. $\angle EHG = \angle AFE = 40°$. $\angle HEG$ must equal $110°$ because there are $180°$ in a triangle. Since $\angle FEG$ is the supplement of $\angle HEG$, $\angle FEG = 70°$.

9. **(C)**

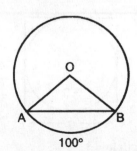

Angle O is a central angle equal to its arc, $100°$. This leaves $80°$ for the other two angles. Since the triangle is isosceles (because the legs are both radii and therefore equal), angle ABO is $40°$.

10. **(A)**
$$d = \sqrt{(5 - (3))^2 + (-5 - 1)^2}$$
$$= \sqrt{(8)^2 + (-6)^2} = \sqrt{64 + 36}$$
$$= \sqrt{100} = 10$$

Exercise 1

1. **(C)** Find the area in square feet and then convert to square yards by dividing by 9. Remember there are 9 square feet in one square yard.

   $(18 \cdot 20) \div 9 = 360 \div 9 = 40$ square yards

2. **(B)** Area of parallelogram $= b \cdot h$

   $$(x+7)(x-7) = 15$$
   $$x^2 - 49 = 15$$
   $$x^2 = 64$$
   $$x = 8$$
   $$\text{Base} = x + 7 = 15$$

3. **(B)** Area of triangle $= \dfrac{1}{2} \cdot b \cdot h$

   Using one leg as base and the other as altitude, the area is $\dfrac{1}{2} \cdot 6 \cdot 8 = 24$. Using the hypotenuse as base and the altitude to the hypotenuse will give the same area.

   $$\frac{1}{2} \cdot 10 \cdot h = 24$$
   $$5h = 24$$
   $$h = 4.8 \therefore \frac{1}{2} \cdot 10 \cdot 4.8 = 24$$

4. **(E)** Area of rhombus $= \dfrac{1}{2} \cdot$ product of diagonals

   $$\text{Area} = \frac{1}{2}(4x)(6x) = \frac{1}{2}(24x^2) = 12x^2$$

5. **(C)**

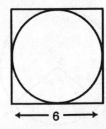

   radius of circle $= 3$

   Area $= \pi r^2 = 9\pi$

Exercise 2

1. **(A)** Area of equilateral triangle $= \dfrac{s^2}{4}\sqrt{3}$

   Therefore, $\dfrac{s^2}{4}$ must equal 16
   $$s^2 = 64$$
   $$s = 8$$
   Perimeter is $8 + 8 + 8 = 24$

2. **(B)** In 4 hours the hour hand moves through one-third of the circumference of the clock.
   $$C = 2\pi r = 2\pi(3) = 6\pi$$
   $$\frac{1}{3} \cdot 6\pi = 2\pi$$

3. **(D)** Compare $2\pi r$ with $2\pi(r+3)$.

   $$2\pi(r+3) = 2\pi r + 6\pi$$

   Circumference was increased by 6π. Trying this with a numerical value for r will give the same result.

4. **(E)** In one revolution, the distance covered is equal to the circumference.

   $$C = 2\pi r = 2\pi(18) = 36\pi \text{ inches}$$

   To change this to feet, divide by 12.

   $$\frac{36\pi}{12} = 3\pi \text{ feet}$$

   In 20 revolutions, the wheel will cover $20(3\pi)$ or 60π feet.

5. **(D)** Area of rectangle $= b \cdot h \qquad = 36$
   Area of square $= s^2 \qquad\qquad = 36$
   Therefore, $s = 6$ and perimeter $= 24$

Exercise 3

1. **(A)**

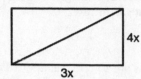

$$14x = 140$$
$$x = 10$$

The rectangle is 30′ by 40′. This is a 3, 4, 5 right triangle, so the diagonal is 50′.

2. **(C)** The altitude in an equilateral triangle is always $\frac{1}{2}$ side $\cdot \sqrt{3}$.

3. **(D)** This is an 8, 15, 17 triangle, making the missing side (3)17, or 51.

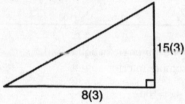

4. **(A)** The diagonal in a square is equal to the side times $\sqrt{2}$. Therefore, the side is 6 and the perimeter is 24.

5. **(C)**

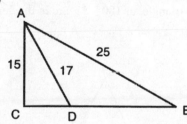

Triangle *ABC* is a 3, 4, 5 triangle with all sides multiplied by 5. Therefore *CB* = 20. Triangle *ACD* is an 8, 15, 17 triangle. Therefore *CD* = 8. *CB* − *CD* = *DB* = 12.

Exercise 4

1. **(A)** Find the midpoint of *AB* by averaging the *x* coordinates and averaging the *y* coordinates.
$$\left(\frac{6+2}{2}, \frac{2+6}{2}\right) = (4,4)$$

2. **(C)** *O* is the midpoint of *AB*.

$$\frac{x+4}{2} = 2 \qquad x+4 = 4, x = 0$$
$$\frac{y+6}{2} = 1 \qquad y+6 = 2, y = -4$$

A is the point (0, −4).

3. **(A)** $d = \sqrt{(8-4)^2 + (6-3)^2} = \sqrt{4^2 + 3^2}$
$$= \sqrt{16+9} = \sqrt{25} = 5$$

4. **(D)** Sketch the triangle and you will see it is a right triangle with legs of 4 and 3.

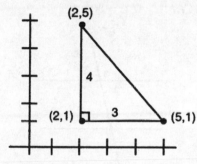

$$\text{Area} = \frac{1}{2} \cdot b \cdot h = \frac{1}{2} \cdot 4 \cdot 3 = 6$$

5. **(A)** Area of a circle = πr^2

$$\pi r^2 = 16\pi \qquad r = 4$$

The point (4, 4) lies at a distance of $\sqrt{(4-0)^2 + (4-0)^2} = \sqrt{32}$ units from (0, 0). All the other points lie 4 units from (0, 0).

Exercise 5

1. **(A)** Angle B = Angle C because of alternate interior angles. Then Angle C = Angle D for the same reason. Therefore, Angle $D = 30°$.

2. **(D)**

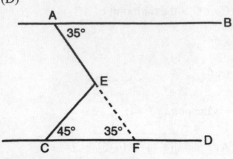

Extend $\overline{AE}$ to F. $\angle A = \angle EFC$

$\angle CEF$ must equal $100°$ because there are $180°$ in a triangle. $\angle AEC$ is supplementary to $\angle CEF$. $\angle AEC = 80°$

3. **(E)**

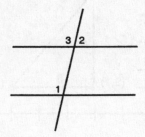

$$\angle 1 = \angle 3$$
$$\angle 2 + \angle 3 = 180°$$
$$\angle 2 = 50°$$

4. **(C)** Since $\angle BEG$ and $\angle EGD$ add to $180°$, halves of these angles must add to $90°$. Triangle EFG contains $180°$, leaving $90°$ for $\angle EFG$.

5. **(C)**

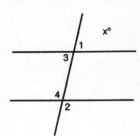

$$\angle 1 = \angle 3$$
$$\angle 2 = \angle 4$$
$$\angle 1 + \angle 2 = \angle 3 + \angle 4$$

But $\angle 3 + \angle 4 = 180°$. Therefore, $\angle 1 + \angle 2 = 180°$

Exercise 6

1. **(D)** Represent the angles as x, $5x$, and $6x$. They must add to $180°$.
$$12x = 180$$
$$x = 15$$

The angles are $15°$, $75°$, and $90°$. Thus, it is a right triangle.

2. **(D)** There are $130°$ left to be split evenly between the base angles (the base angles must be equal). Each one must be $65°$.

3. **(E)**

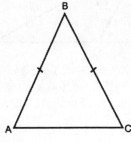

The exterior angle is equal to the sum of the two remote interior angles.

$$4x = 100$$
$$x = 25$$
$$\text{Angle } A = 3x = 75°$$

4. **(D)** The other base angle is also x. These two base angles add to $2x$. The remaining degrees of the triangle, or $180 - 2x$, are in the vertex angle.

5. **(E)**

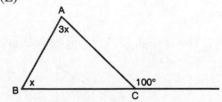

$$\angle A = \angle C$$
$$4x - 30 = 2x + 10$$
$$2x = 40$$
$$x = 20$$

$\angle A$ and $\angle C$ are each $50°$, leaving $80°$ for $\angle B$.

Exercise 7

1. **(C)** A hexagon has 6 sides. Sum = $(n - 2)$ 180 = $4(180) = 720$

2. **(D)** Opposite sides of a parallelogram are congruent, so $AB = CD$.

$$x + 4 = 2x - 16$$
$$20 = x$$
$$AD = BC = x - 6 = 14$$

3. **(B)** $AB = CD$

$$x + 8 = 4x - 4$$
$$12 = 3x$$
$$x = 4$$
$$AB = 12 \qquad BC = 12 \qquad CD = 12$$

If all sides are congruent, it must be a rhombus. Additional properties would be needed to make it a square.

4. **(B)** A rhombus has 4 sides. Sum = $(n - 2)$ 180 = $2(180) = 360$

5. **(C)** Rectangles and rhombuses are both types of parallelograms but do not share the same special properties. A square is both a rectangle and a rhombus with *added* properties.

Exercise 8

1. **(C)** Tangent segments drawn to a circle from the same external point are congruent. If $CE = 5$, then $CF = 5$, leaving 7 for BF. Therefore BD is also 7. If $AE = 2$, then $AD = 2$.

$$BD + DA = BA = 9$$

2. **(D)** Angle O is a central angle equal to its arc, $40°$. This leaves $140°$ for the other two angles. Since the triangle is isosceles, because the legs are equal radii, each angle is $70°$.

3. **(E)** The remaining arc is $120°$. The inscribed angle x is $\frac{1}{2}$ its intercepted arc.

4. **(A)**
$$50° = \frac{1}{2}\left(40° + AC\right)$$
$$100° = 40° + AC$$
$$60° = AC$$

5. **(D)** An angle outside the circle is $\frac{1}{2}$ the difference of its intercepted arcs.

Exercise 9

1. **(D)** There are 6 equal squares in the surface area of a cube. Each square will have an area of $\frac{96}{6}$ or 16. Each edge is 4.
$$V = e^3 = 4^3 = 64$$

2. **(C)** $V = \pi r^2 h = \frac{22}{7} \cdot 49 \cdot 10 = 1540$ cubic inches

   Divide by 231 to find gallons.

3. **(B)** $V = \pi r^2 h = \frac{22}{7} \cdot 9 \cdot 14 = 396$ cubic inches

   Divide by 9 to find minutes.

4. **(B)** $V = l \cdot w \cdot h = 10 \cdot 8 \cdot 4 = 320$ cubic inches

   Each small cube = $4^3 = 64$ cubic inches. Therefore it will require 5 cubes.

5. **(A)** Change 16 inches to $1\frac{1}{3}$ feet.
$$V = 6 \cdot 5 \cdot 1\frac{1}{3} = 40 \text{ cubic feet when full.}$$
$$\frac{5}{8} \cdot 40 = 25$$

Exercise 10

1. **(E)** If the radius is multiplied by 3, the area is multiplied by 3^2 or 9.

2. **(D)** If the dimensions are all doubled, the area is multiplied by 2^2 or 4. If the new area is 4 times as great as the original area, is has been *increased* by 300%.

3. **(A)** If the area ratio is 9 : 1, the linear ratio is 3 : 1. Therefore, the larger radius is 3 times the smaller radius.

4. **(B)** Ratio of circumferences is the same as ratio of radii, but the area ratio is the square of this.

5. **(C)** We must take the cube root of the volume ratio to find the linear ratio. This becomes much easier if you simplify the ratio first.
$$\frac{250}{128} = \frac{125}{64}$$
The linear ratio is then 5 : 4.
$$\frac{5}{4} = \frac{25}{x}$$
$$5x = 100$$
$$x = 20$$

Retest

1. (C)   Area of trapezoid = $\frac{1}{2}h(b_1+b_2)$

    Area $= \frac{1}{2}\cdot 3(10+12)=33$

2. (A)   Arca of circle = $\pi r^2 = 16\pi$

    Therefore, $r^2 = 16$ or $r = 4$

    Circumference of circle $= 2\pi r = 2\pi\,(4) = 8\pi$

3. (D)   The side of a square is equal to the diagonal times $\frac{\sqrt{2}}{2}$. Therefore, the side is $4\sqrt{2}$ and the perimeter is $16\sqrt{2}$.

4. (E) $d = \sqrt{(7-4)^2 + (-7-(-3))^2}$

    $= \sqrt{(3)^2 + (-4)^2} = \sqrt{9+16}$

    $= \sqrt{25} = 5$

5. (D)

    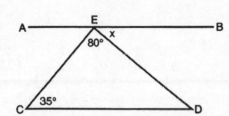

    $\angle CDE$ must equal 65° because there are 180° in a triangle. Since $\overline{AB}$ is parallel to $\overline{CD}$, $\angle x = \angle CDE = 65°$.

6. (C)   Represent the angles as x, x, and $2x$. They must add to 180°.

    $4x = 180$

    $x = 45$

    Therefore, the largest angle is $2x = 2(45°) = 90°$.

7. (B)   A pentagon has 5 sides. Sum $(n-2)180 = 3(180) = 540°$

    In a regular pentagon, all the angles are equal. Therefore, each angle $= \frac{540}{5} = 108°$.

8. (D)

    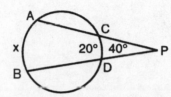

    An angle outside the circle is $\frac{1}{2}$ the difference of its intercepted arcs.

    $40 = \frac{1}{2}(x-20)$

    $80 = x - 20$

    $100 = x$

9. (D) $V = 1 \cdot w \cdot h = 2 \cdot 6 \cdot 18 = 216$

    The volume of a cube is equal to thc cube of an edge.

    $V = e^3$

    $216 = e^3$

    $6 = e$

10. (B)   If the volume ratio is 8 : 1, the linear ratio is 2 : 1, and the area ratio is the square of this, or 4:1.

Inequalities

14

DIAGNOSTIC TEST

Directions: Work out each problem. Circle the letter that appears before your answer.

Answers are at the end of the chapter.

1. If $4x < 6$, then

   (A) $x = 1.5$

   (B) $x < \dfrac{2}{3}$

   (C) $x > \dfrac{2}{3}$

   (D) $x < \dfrac{3}{2}$

   (E) $x > \dfrac{3}{2}$

2. a and b are positive numbers. If $a = b$ and $c > d$, then

   (A) $a + c < b + d$

   (B) $a + c > b + d$

   (C) $a - c > b - d$

   (D) $ac < bd$

   (E) $a + c < b - d$

3. Which value of x will make the following expression true?

$$\frac{3}{5} < \frac{x}{10} < \frac{4}{5}$$

   (A) 5

   (B) 6

   (C) 7

   (D) 8

   (E) 9

4. In triangle ABC, $AB = AC$ and $EC < DB$. Then

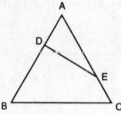

   (A) $DB < AE$

   (B) $DB < AD$

   (C) $AD > AE$

   (D) $AD < AE$

   (E) $AD > EC$

5. In triangle ABC, $\angle 1 > \angle 2$ and $\angle 2 > \angle 3$. Then

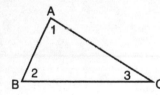

   (A) $AC < AB$

   (B) $AC > BC$

   (C) $BC > AC$

   (D) $BC < AB$

   (E) $\angle 3 > \angle 1$

6. If point C lies between A and B on line segment AB, which of the following is always true?

   (A) $AC = CB$

   (B) $AC > CB$

   (C) $CB > AC$

   (D) $AB < AC + CB$

   (E) $AB = CB + AC$

7. If AC is perpendicular to BD, which of the following is always true?

I. $AC = BC$
II. $AC < AB$
III. $AB > AD$

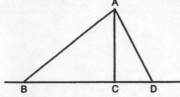

(A) I only
(B) II and III only
(C) II only
(D) III only
(E) I and II only

8. If $x < 0$ and $y > 0$, which of the following is always true?

(A) $x + y > 0$
(B) $x + y < 0$
(C) $y - x < 0$
(D) $x - y < 0$
(E) $2x > y$

9. In triangle ABC, BC is extended to D. If $\angle A = 50°$ and $\angle ACD = 120°$, then

(A) $BC > AB$
(B) $AC > AB$
(C) $BC > AC$
(D) $AB > AC$
(E) $\angle B < \angle A$

10. In right triangle ABC, $\angle A < \angle B$ and $\angle B < \angle C$. Then

(A) $\angle A > 45°$
(B) $\angle B = 90°$
(C) $\angle B > 90°$
(D) $\angle C = 90°$
(E) $\angle C > 90°$

1. ALGEBRAIC INEQUALITIES

Algebraic inequality statements are solved in the same manner as equations. However, do not forget that whenever you multiply or divide by a negative number, the order of the inequality, that is, the inequality symbol must be reversed. In reading the inequality symbol, remember that it points to the smaller quantity. $a < b$ is read a is less than b. $a > b$ is read a is greater than b.

Example:

Solve for x: $12 - 4x < 8$

Solution:

Add -12 to each side.
$-4x < -4$
Divide by -4, remembering to reverse the inequality sign.
$x > 1$

Example:

$6x + 5 > 7x + 10$

Solution:

Collect all the terms containing x on the left side of the equation and all numerical terms on the right. As with equations, remember that if a term comes from one side of the inequality to the other, that term changes sign.
$-x > 5$
Divide (or multiply) by -1.
$x < -5$

Exercise 1

Work out each problem. Circle the letter that appears before your answer.

1. Solve for x: $8x < 5(2x + 4)$

    (A) $x > -10$
    (B) $x < -10$
    (C) $x > 10$
    (D) $x < 10$
    (E) $x < 18$

2. Solve for x: $6x + 2 - 8x < 14$

    (A) $x = 6$
    (B) $x = -6$
    (C) $x > -6$
    (D) $x < -6$
    (E) $x > 6$

3. A number increased by 10 is greater than 50. What numbers satisfy this condition?

    (A) $x > 60$
    (B) $x < 60$
    (C) $x > -40$
    (D) $x < 40$
    (E) $x > 40$

4. Solve for x: $-.4x < 4$

    (A) $x > -10$
    (B) $x > 10$
    (C) $x < 8$
    (D) $x < -10$
    (E) $x < 36$

5. Solve for x: $.03n > -.18$

    (A) $n < -.6$
    (B) $n > .6$
    (C) $n > 6$
    (D) $n > -6$
    (E) $n < -6$

6. Solve for b: $15b < 10$

    (A) $b < \dfrac{3}{2}$

    (B) $b > \dfrac{3}{2}$

    (C) $b < -\dfrac{3}{2}$

    (D) $b < \dfrac{2}{3}$

    (E) $b > \dfrac{2}{3}$

7. If $x^2 < 4$, then

    (A) $x > 2$
    (B) $x < 2$
    (C) $x > -2$
    (D) $-2 < x < 2$
    (E) $-2 \leq x \leq 2$

8. Solve for n: $n + 4.3 < 2.7$

    (A) $n > 1.6$
    (B) $n > -1.6$
    (C) $n < 1.6$
    (D) $n < -1.6$
    (E) $n = 1.6$

9. If $x < 0$ and $y < 0$, which of the following is always true?

    (A) $x + y > 0$
    (B) $xy < 0$
    (C) $x - y > 0$
    (D) $x + y < 0$
    (E) $x = y$

10. If $x < 0$ and $y > 0$, which of the following will always be greater than 0?

    (A) $x + y$
    (B) $x - y$
    (C) $\dfrac{x}{y}$
    (D) xy
    (E) $-2x$

2. GEOMETRIC INEQUALITIES

In working with geometric inequalities, certain postulates and theorems should be reviewed.

A. If unequal quantities are added to unequal quantities of the same order, the sums are unequal in the same order.

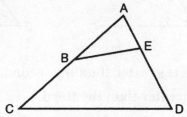

If $AB > AE$ and
(+) $BC > ED$ then
  $AC > AD$

B. If equal quantities are added to unequal quantities, the sums are unequal in the same order.

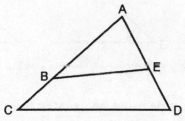

  $AB > AE$ and
(+) $BC = ED$ then
  $AC > AD$

C. If equal quantities are subtracted from unequal quantities, the differences are unequal in the same order.

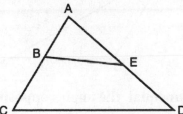

If $AC > AD$ and
(−) $BC = ED$ then
  $AB > AE$

D. If unequal quantities are subtracted from equal quantities, the results are unequal in the *opposite* order.

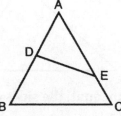

  $AB = AC$
(−) $AD < AE$
  $DB > EC$

E. Doubles of unequals are unequal in the same order.

M is the midpoint of AB

N is the midpoint of CD

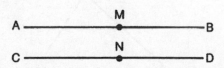

$AM > CN$
Therefore, $AB > CD$

F. Halves of unequals are unequal in the same order.

$\angle ABC > \angle DEF$

$\overline{BG}$ bisects $\angle ABC$

$\overline{EH}$ bisects $\angle DEF$

Therefore, $\angle 1 > \angle 2$

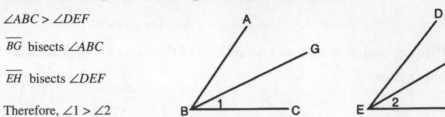

G. If the first of three quantities is greater than the second, and the second is greater than the third, then the first is greater than the third.

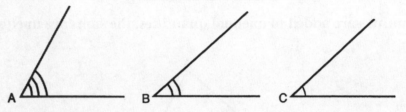

If $\angle A > \angle B$ and $\angle B > \angle C$, then $\angle A > \angle C$.

H. The sum of two sides of a triangle must be greater than the third side.

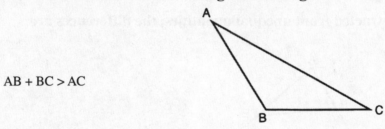

$AB + BC > AC$

I. If two sides of a triangle are unequal, the angles opposite are unequal, with the larger angle opposite the larger side.

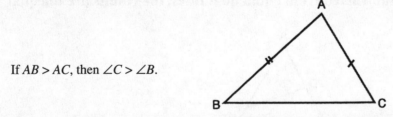

If $AB > AC$, then $\angle C > \angle B$.

J. If two angles of a triangle are unequal, the sides opposite these angles are unequal, with the larger side opposite the larger angle.

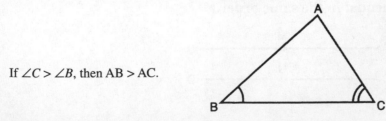

If $\angle C > \angle B$, then $AB > AC$.

K. An exterior angle of a triangle is greater than either remote interior angle.

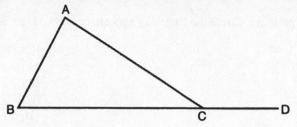

∠ACD > ∠B and ∠ACD > ∠A

Exercise 2

Work out each problem. Circle the letter that appears before your answer.

1. Which of the following statements is true regarding triangle *ABC?*

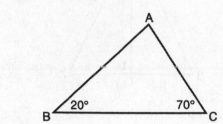

   (A) *AC > AB*
   (B) *AB > BC*
   (C) *AC > BC*
   (D) *BC > AB*
   (E) *BC > AB + AC*

2. In triangle *RST*, *RS = ST*. If *P* is any point on *RS*, which of the following statements is always true?

   (A) *PT < PR*
   (B) *PT > PR*
   (C) *PT = PR*
   (D) *PT = ½ PR*
   (E) *PT ≤ PR*

3. If ∠A > ∠C and ∠ABD = 120°, then

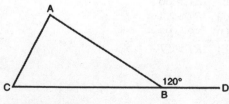

   (A) *AC < AB*
   (B) *BC < AB*
   (C) ∠C > ∠ABC
   (D) *BC > AC*
   (E) ∠ABC > ∠A

4. If *AB ⊥ CD* and ∠1 > ∠4, then

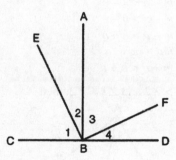

   (A) ∠1 > ∠2
   (B) ∠4 > ∠3
   (C) ∠2 > ∠3
   (D) ∠2 < ∠3
   (E) ∠2 < ∠4

5. Which of the following sets of numbers could be the sides of a triangle?

   (A) 1, 2, 3
   (B) 2, 2, 4
   (C) 3, 3, 6
   (D) 1, 1.5, 2
   (E) 5, 6, 12

RETEST

Work out each problem. Circle the letter that appears before your answer.

1. If $2x > -5$, then

   (A) $x > \dfrac{5}{2}$

   (B) $x > -\dfrac{5}{2}$

   (C) $x > -\dfrac{2}{5}$

   (D) $x < \dfrac{5}{2}$

   (E) $x < -\dfrac{5}{2}$

2. $m, n > 0$. If $m = n$ and $p < q$, then

   (A) $m - p < n - q$
   (B) $p - m > q - n$
   (C) $m - p > n - q$
   (D) $mp > nq$
   (E) $m + q < n + p$

3. If $\angle 3 > \angle 2$ and $\angle 1 = \angle 2$, then

   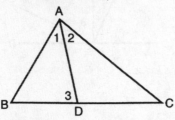

   (A) $AB > BD$
   (B) $AB < BD$
   (C) $DC = BD$
   (D) $AD > BD$
   (E) $AB < AC$

4. If $\angle 1 > \angle 2$ and $\angle 2 > \angle 3$, then

   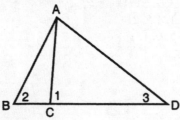

   (A) $AB > AD$
   (B) $AC > AD$
   (C) $AC < CD$
   (D) $AD > AC$
   (E) $AB > BC$

5. If $\dfrac{x}{2} > 6$, then

   (A) $x > 3$
   (B) $x < 3$
   (C) $x > 12$
   (D) $x < 12$
   (E) $x > -12$

6. If $AB = AC$ and $\angle 1 > \angle B$, then

   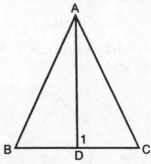

   (A) $\angle B > \angle C$
   (B) $\angle 1 > \angle C$
   (C) $BD > AD$
   (D) $AB > AD$
   (E) $\angle ADC > \angle ADB$

7. Which of the following sets of numbers may be used as the sides of a triangle?

   (A) 7, 8, 9
   (B) 3, 5, 8
   (C) 8, 5, 2
   (D) 3, 10, 6
   (E) 4, 5, 10

8. In isosceles triangle RST, $RS = ST$. If A is the midpoint of RS and B is the midpoint of ST, then

   (A) $SA > ST$
   (B) $BT > BS$
   (C) $BT = SA$
   (D) $SR > RT$
   (E) $RT > ST$

9. If $x > 0$ and $y < 0$, which of the following is always true?

   (A) $x - y > y - x$
   (B) $x + y > 0$
   (C) $xy > 0$
   (D) $y > x$
   (E) $x - y < 0$

10. In triangle ABC, AD is the altitude to BC. Then

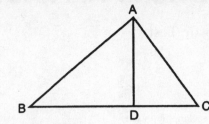

   (A) $AD > DC$
   (B) $AD < BD$
   (C) $AD > AC$
   (D) $BD > DC$
   (E) $AB > BD$

SOLUTIONS TO PRACTICE EXERCISES

Diagnostic Test

1. (D) $4x < 6$

    $x < \dfrac{6}{4}$

    Simplify to $x < \dfrac{3}{2}$

2. (B)   If equal quantities are added to unequal quantities, the sums are unequal in the same order.

    $c > d$

    $\underline{(+)\,a = b}$

    $a + c > b + d$

3. (C) $\dfrac{3}{5} < \dfrac{x}{10} < \dfrac{4}{5}$

    Multiply through by 10.

    $6 < x < 8$ or x must be between 6 and 8.

4. (D)

    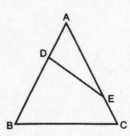

    If unequal quantities are subtracted from equal quantities, the results are unequal in the opposite order.

    $AC = AB$

    $\underline{(-)\,EC < DB}$

    $AE > AD$ or $AD < AE$

5. (C)   If two angles of a triangle are unequal, the sides opposite these angles are unequal, with the larger side opposite the larger angle.

    Since $\angle 1 > \angle 2$, $BC > AC$.

6. (E)

    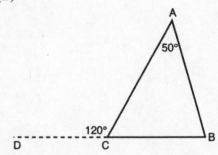

    $AB = CB + AC$

7. (C) In right triangle ACB, the longest side is the hypotenuse AB. Therefore, side AC is less than AB.

8. (D) A positive subtracted from a negative is always negative.

9. (B)

    $AB = CB + AC$

    $\angle ACB$ is the supplement of $\angle ACD$. Therefore, $\angle ACB = 60°$. $\angle ABC$ must equal 70° because there are 180° in a triangle. Since $\angle ABC$ is the largest angle in the triangle, AC must be the longest side. Therefore, $AC > AB$.

10. (D)   In a right triangle, the largest angle is the right angle. Since $\angle C$ is the largest angle, $\angle C = 90°$.

Exercise 1

1. (A) $8x < 10x + 20$
    $-2x < 20$
    $x > -10$

2. (C) $-2x < 12$
    $x > -6$

3. (E) $x + 10 > 50$
    $x > 40$

4. (A) $-.4x < 4$

    Multiply by 10 to remove decimals.
    $-4x < 40$
    $x > -10$

5. (D) $.03n > -.18$

    Multiply by 100
    $3n > -18$
    $n > -6$

6. (D) Divide by 15
    $$b < \frac{10}{15}$$
    Simplify to $b < \frac{2}{3}$

7. (D) x must be less than 2, but can go no lower than -2, as $(-3)^2$ would be greater than 4.

8. (D) $n + 4.3 < 2.7$

    Subtract 4.3 from each side.
    $n < -1.6$

9. (D) When two negative numbers are added, their sum will be negative.

10. (E) The product of two negative numbers is positive.

Exercise 2

1. (D) Angle A will contain 90°, which is the largest angle of the triangle. The sides from largest to smallest will be *BC*, *AB*, *AC*.

2. (B) Since $\angle SRT = \angle STR$, $\angle SRT$ will have to be greater than $\angle PTR$. Therefore, $PT > PR$ in triangle *PRT*.

3. (D) Angle $ABC = 60°$. Since there are 120° left for $\angle A$ and $\angle C$ together and, also $\angle A > \angle C$, then $\angle A$ must contain more than half of 120° and $\angle C$ must contain less than half of 120°. This makes $\angle A$ the largest angle of the triangle. The sides in order from largest to smallest are *BC*, *AC*, *AB*.

4. (D) $\angle ABC = \angle ABD$ as they are both right angles. If $\angle 1 > \angle 4$, then $\angle 2$ will be less than $\angle 3$ because we are subtracting unequal quantities ($\angle 1$ and $\angle 4$) from equal quantities ($\angle ABC$ and $\angle ABD$).

5. (D) The sum of any two sides (always try the shortest two) must be greater than the third side.

Retest

1. (B) $2x > -5$

$$x > -\frac{5}{2}$$

2. (C)   If unequal quantities are subtracted from equal quantities, the differences are unequal in the opposite order.

$$m = n$$
$$(-)p < q$$
$$m - p > n - q$$

3. (A)   Since $\angle 3 > \angle 2$ and $\angle 1 = \angle 2$, $\angle 3 > \angle 1$. If two angles of a triangle are unequal, the sides opposite these angles are unequal, with the larger side opposite the larger angle. Therefore, $AB > BD$.

4. (D)   Since $\angle 1 > \angle 2$ and $\angle 2 > \angle 3$, $\angle 1 > \angle 3$. In triangle ACD side AD is larger than side AC, since AD is opposite the larger angle.

5. (C) $\dfrac{x}{2} > 6$

$$x > 12$$

6. (B)   If two sides of a triangle are equal, the angles opposite them are equal. Therefore $\angle C = \angle B$. Since $\angle 1 > \angle B$, $\angle 1 > \angle C$.

7. (A)   The sum of any two sides (always try the shortest two) must be greater than the third side.

8. (C)

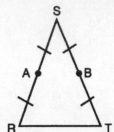

$BT = \dfrac{1}{2}\ ST$ and $SA = \dfrac{1}{2}\ SR$. Since $ST = SR$, $BT = SA$.

9. (A)   A positive minus a negative is always greater than a negative minus a positive.

10. (E)   In right triangle ADB, the longest side is the hypotenuse AB. Therefore, $AB > BD$.

Numbers and Operations, Algebra, and Functions

15

DIAGNOSTIC TEST

Directions: Answer the following 10 questions, limiting your time to 15 minutes. Note that question 1 is a *grid-in* question, in which you provide the numerical solution. (All other questions are in multiple-choice format.)

Answers are at the end of the chapter.

1. The population of Urbanville has always doubled every five years. Urbanville's current population is 25,600. What was its population 20 years ago?

2. Which of the following describes the union of the factors of 15, the factors of 30, and the factors of 75?

   (A) The factors of 15
   (B) The factors of 30
   (C) The factors of 45
   (D) The factors of 75
   (E) None of the above

3. $|-1 - 2| - |5 - 6| - |-3 + 4| =$

   (A) −5
   (B) −3
   (C) 1
   (D) 3
   (E) 5

4. For all $x \neq 0$ and $y \neq 0$, $\dfrac{x^6 y^3 y}{y^6 x^3 x}$ is equivalent to:

   (A) $\dfrac{y^2}{x^2}$
   (B) xy
   (C) 1
   (D) $\dfrac{x^2}{y^2}$
   (E) $\dfrac{x^3}{y^3}$

5. If $f(x) = x + 1$, then $\dfrac{1}{f(x)} \times f\left(\dfrac{1}{x}\right) =$

   (A) 1
   (B) $\dfrac{1}{x}$
   (C) x
   (D) $\dfrac{x}{x+1}$
   (E) x^2

6. If the domain of $f(x) = \dfrac{x}{5^{-x}}$ is the set $\{-2, -1, 0, 2\}$, then $f(x)$ CANNOT equal

   (A) $-\dfrac{2}{25}$
   (B) $-\dfrac{1}{5}$
   (C) 0
   (D) 5
   (E) 50

7. Which of the following equations defines a function containing the (x,y) pairs $(-1,-1)$ and $(-\frac{1}{2},0)$?

   (A) $y = 3x + 2$
   (B) $y = 2x + 1$
   (C) $y = 6x + 5$
   (D) $y = -4x - 2$
   (E) $y = 4x + 3$

8. The figure below shows the graph of a linear function on the *xy*-plane.

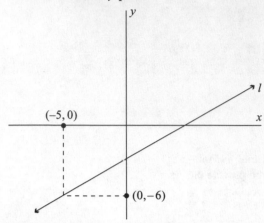

If the *x*-intercept of line *l* is 4, what is the slope of *l* ?

(A) $\frac{2}{3}$

(B) $\frac{3}{4}$

(C) $\frac{5}{6}$

(D) $\frac{6}{5}$

(E) Not enough information to answer the question is given.

9. The figure below shows a parabola in the *xy*-plane.

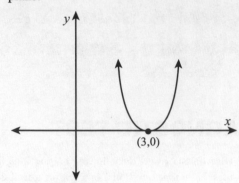

Which of the following equations does the graph best represent?

(A) $y = -x^2 + 6x - 9$

(B) $y = x^2 - 2x + 6$

(C) $y = \frac{2}{3}x^2 - 4x + 6$

(D) $y = -x^2 + x - 3$

(E) $y = x^2 + 3x + 9$

10. Which of the following best describes the relationship between the graph of $y = \frac{2}{x^2}$ and the graph of $x = \frac{2}{y^2}$ in the *xy*-plane?

(A) Mirror images symmetrical about the *x*-axis

(B) Mirror images symmetrical about the *y*-axis

(C) Mirror images symmetrical about the line of the equation $x = y$

(D) Mirror images symmetrical about the line of the equation $x = -y$

(E) None of the above

1. SEQUENCES INVOLVING EXPONENTIAL GROWTH (GEOMETRIC SEQUENCES)

In a sequence of terms involving exponential growth, which the testing service also calls a *geometric sequence*, there is a constant ratio between consecutive terms. In other words, each successive term is the same multiple of the preceding one. For example, in the sequence 2, 4, 8, 16, 32, . . . , notice that you multipy each term by 2 to obtain the next term, and so the constant ratio (multiple) is 2.

To solve problems involving geometric sequence, you can apply the following standard equation:

$$a \cdot r^{(n-1)} = T$$

In this equation:

The variable a is the value of the first term in the sequence

The variable r is the constant ratio (multiple)

The variable n is the position number of any particular term in the sequence

The variable T is the value of term n

If you know the values of any three of the four variables in this standard equation, then you can solve for the fourth one. (On the SAT, geometric sequence problems generally ask for the value of either a or T.)

Example (solving for T when a and r are given):

The first term of a geometric sequence is 2, and the constant multiple is 3. Find the second, third, and fourth terms.

Solution:

2nd term $(T) = 2 \cdot 3^{(2-1)} = 2 \cdot 3^1 = 6$

3rd term $(T) = 2 \cdot 3^{(3-1)} = 2 \cdot 3^2 = 2 \cdot 9 = 18$

4th term $(T) = 2 \cdot 3^{(4-1)} = 2 \cdot 3^3 = 2 \cdot 27 = 54$

To solve for T when a and r are given, as an alternative to applying the standard equation, you can multiply a by $r^{(n-1)}$ times. Given $a = 2$ and $r = 3$:

2nd term $(T) = 2 \cdot 3 = 6$

3rd term $(T) = 2 \cdot 3 = 6 \cdot 3 = 18$

4th term $(T) = 2 \cdot 3 = 6 \cdot 3 = 18 \cdot 3 = 54$

NOTE: Using the alternative method, you may wish to use your calculator to find T if a and/or r are large numbers.

Example (solving for a when r and T are given):

The fifth term of a geometric sequence is 768, and the constant multiple is 4. Find the 1st term (a).

Solution:

$$a \times 4^{(5-1)} = 768$$
$$a \times 4^4 = 768$$
$$a \times 256 = 768$$
$$a = \frac{768}{256}$$
$$a = 3$$

Example (solving for T when a and another term in the sequence are given):

To find a particular term (T) in a geometric sequence when the first term and another term are given, first determine the constant ratio (r), and then solve for T. For example, assume that the first and sixth terms of a geometric sequence are 2 and 2048, respectively. To find the value of the fourth term, first apply the standard equation to determine r:

Solution:

$$2 \times r^{(6-1)} = 2048$$
$$2 \times r^5 = 2048$$
$$r^5 = \frac{2048}{2}$$
$$r^5 = 1024$$
$$r = \sqrt[5]{1024}$$
$$r = 4$$

The constant ratio is 4. Next, in the standard equation, let $a = 2$, $r = 4$, and $n = 4$, and then solve for T:

$$2 \times 4^{(4-1)} = T$$
$$2 \times 4^3 = T$$
$$2 \times 64 = T$$
$$128 = T$$

The fourth term in the sequence is 128.

Exercise 1

Work out each problem. For questions 1–3, circle the letter that appears before your answer. Questions 4 and 5 are grid-in questions.

1. On January 1, 1950, a farmer bought a certain parcel of land for $1,500. Since then, the land has doubled in value every 12 years. At this rate, what will the value of the land be on January 1, 2010?

    (A) $7,500
    (B) $9,000
    (C) $16,000
    (D) $24,000
    (E) $48,000

2. A certain type of cancer cell divides into two cells every four seconds. How many cells are observable 32 seconds after observing a total of four cells?

    (A) 1,024
    (B) 2,048
    (C) 4,096
    (D) 5,512
    (E) 8,192

3. The seventh term of a geometric sequence with constant ratio 2 is 448. What is the first term of the sequence?

    (A) 6
    (B) 7
    (C) 8
    (D) 9
    (E) 11

4. Three years after an art collector purchases a certain painting, the value of the painting is $2,700. If the painting increased in value by an average of 50 percent per year over the three year period, how much did the collector pay for the painting, in dollars?

5. What is the second term in a geometric series with first term 3 and third term 147?

2. SETS (UNION, INTERSECTION, ELEMENTS)

A *set* is simply a collection of elements; elements in a set are also referred to as the "members" of the set. An SAT problem involving sets might ask you to recognize either the union or the intersection of two (or more) sets of numbers.

The *union* of two sets is the set of all members of either or both sets. For example, the union of the set of all negative integers and the set of all non-negative integers is the set of all integers. The *intersection* of two sets is the set of all common members – in other words, members of *both* sets. For example, the intersection of the set of integers less than 11 and the set of integers greater than 4 but less than 15 is the following set of six consecutive integers: {5,6,7,8,9,10}.

On the new SAT, a problem involving either the union or intersection of sets might apply any of the following concepts: the real number line, integers, multiples, factors (including prime factors), divisibility, or counting.

Example:

Set A is the set of all positive multiples of 3, and set B is the set of all positive multiples of 6. What is the union and intersection of the two sets?

Solution:

The union of sets A and B is the set of all postitive multiples of 3.
The intersection of sets A and B is the set of all postitive multiples of 6.

Exercise 2

Work out each problem. Note that question 2 is a grid-in question. For all other questions, circle the letter that appears before your answer.

1. Which of the following describes the union of the set of integers less than 20 and the set of integers greater than 10?

    (A) Integers 10 through 20
    (B) All integers greater than 10 but less than 20
    (C) All integers less than 10 and all integers greater than 20
    (D) No integers
    (E) All integers

2. Set A consists of the positive factors of 24, and set B consists of the positive factors of 18. The intersection of sets A and B is a set containing how many members?

3. The union of sets X and Y is a set that contains exactly two members. Which of the following pairs of sets could be sets X and Y ?

    (A) The prime factors of 15; the prime factors of 30
    (B) The prime factors of 14; the prime factors of 51
    (C) The prime factors of 19; the prime factors of 38
    (D) The prime factors of 22; the prime factors of 25
    (E) The prime factors of 39; the prime factors of 52

4. The set of all multiples of 10 could be the intersection of which of the following pairs of sets?

    (A) The set of all multiples of $\frac{5}{2}$; the set of all multiples of 2
    (B) The set of all multiples of $\frac{3}{5}$; the set of all multiples of 5
    (C) The set of all multiples of $\frac{3}{2}$; the set of all multiples of 10
    (D) The set of all multiples of $\frac{3}{4}$; the set of all multiples of 2
    (E) The set of all multiples of $\frac{5}{2}$; the set of all multiples of 4

5. For all real numbers x, sets P, Q, and R are defined as follows:

    $P:\{x \geq -10\}$

    $Q:\{x \geq 10\}$

    $R:\{|x| \leq 10\}$

    Which of the following indicates the intersection of sets P, Q, and R ?

    (A) $x =$ any real number
    (B) $x \geq -10$
    (C) $x \geq 10$
    (D) $x = 10$
    (E) $-10 \leq x \leq 10$

3. ABSOLUTE VALUE

The *absolute value* of a real number refers to the number's distance from zero (the origin) on the real-number line. The absolute value of x is indicated as $|x|$. The absolute value of a negative number always has a positive value.

Example:

$|-2 - 3| - |2 - 3| =$
(A) -2
(B) -1
(C) 0
(D) 1
(E) 4

Solution:

The correct answer is (E). $|-2 - 3| = |-5| = 5$, and $|2 - 3| = |-1| = 1$. Performing subtraction: $5 - 1 = 4$.

The concept of absolute value can be incorporated into many different types of problems on the new SAT, including those involving algebraic expressions, equations, and inequalities, as well as problems involving functional notation and the graphs of functions.

Exercise 3

Work out each problem. Circle the letter that appears before your answer.

1. $|7 - 2| - |2 - 7| =$
    (A) -14
    (B) -9
    (C) -5
    (D) 0
    (E) 10

2. For all integers a and b, where $b \neq 0$, subtracting b from a must result in a positive integer if:
    (A) $|a - b|$ is a positive integer
    (B) $\left(\dfrac{a}{b}\right)$ is a positive integer
    (C) $(b - a)$ is a negative integer
    (D) $(a + b)$ is a positive integer
    (E) (ab) is a positive integer

3. What is the complete solution set for the inequality $|x - 3| > 4$?
    (A) $x > -1$
    (B) $x > 7$
    (C) $-1 < x < 7$
    (D) $x < -7, x > 7$
    (E) $x < -1, x > 7$

4. The figure below shows the graph of a certain equation in the xy-plane.

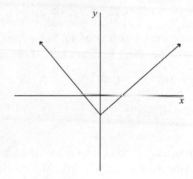

    Which of the following could be the equation?
    (A) $x = |y| - 1$
    (B) $y = |x| - 1$
    (C) $|y| = x - 1$
    (D) $y = x + 1$
    (E) $|x| = y - 1$

5. If $f(x) = |\frac{1}{x} - 3| - x$, then $f\left(\frac{1}{2}\right) =$
    (A) -1
    (B) $-\dfrac{1}{2}$
    (C) 0
    (D) $\dfrac{1}{2}$
    (E) 1

4. EXPONENTS (POWERS)

An *exponent*, or *power*, refers to the number of times that a number (referred to as the *base* number) is multiplied by itself, plus 1. In the number 2^3, the base number is 2 and the exponent is 3. To calculate the value of 2^3, you multiply 2 by itself twice: $2^3 = 2 \cdot 2 \cdot 2 = 8$. In the number $\left(\frac{2}{3}\right)^4$, the base number is $\frac{2}{3}$ and the exponent is 4. To calculate the value of $\left(\frac{2}{3}\right)^4$, you multiply $\frac{2}{3}$ by itself three times: $\left(\frac{2}{3}\right)^4 = \frac{2}{3} \times \frac{2}{3} \times \frac{2}{3} \times \frac{2}{3} = \frac{16}{81}$.

An SAT problem might require you to combine two or more terms that contain exponents. Whether you can you combine base numbers—using addition, subtraction, multiplication, or division—*before* applying exponents to the numbers depends on which operation you're performing. When you add or subtract terms, you cannot combine base numbers or exponents:

$$a^x + b^x \neq (a + b)^x$$
$$a^x - b^x \neq (a - b)^x$$

Example:

If $x = -2$, then $x^5 - x^2 - x =$

(A) 26
(B) 4
(C) −34
(D) −58
(E) −70

Solution:

The correct answer is (C). You cannot combine exponents here, even though the base number is the same in all three terms. Instead, you need to apply each exponent, in turn, to the base number, then subtract:

$$x^5 - x^2 - x = (-2)^5 - (-2)^2 - (-2) = -32 - 4 + 2 = -34$$

There are two rules you need to know for combining exponents by multiplication or division. First, you can combine base numbers first, but only if the exponents are the same:

$$a^x \cdot b^x = (ab)^x$$

$$\frac{a^x}{b^x} = \left(\frac{a}{b}\right)^x$$

Second, you can combine exponents first, but only if the base numbers are the same. When multiplying these terms, add the exponents. When dividing them, subtract the denominator exponent from the numerator exponent:

$$a^x \cdot a^y = a^{(x + y)}$$

$$\frac{a^x}{a^y} = a^{(x-y)}$$

When the same base number (or term) appears in both the numerator and denominator of a fraction, you can factor out, or cancel, the number of powers common to both.

Example:

Which of the following is a simplified version of $\dfrac{x^2 y^3}{x^3 y^2}$?

(A) $\dfrac{y}{x}$

(B) $\dfrac{x}{y}$

(C) $\dfrac{1}{xy}$

(D) 1

(E) $x^5 y^5$

Solution:

The correct answer is (A). The simplest approach to this problem is to cancel, or factor out, x^2 and y^2 from numerator and denominator. This leaves you with x^1 in the denominator and y^1 in the denominator.

You should also know how to raise exponential numbers to powers, as well as how to raise base numbers to negative and fractional exponents. To raise an exponential number to a power, multiply exponents together:

$$\left(a^x\right)^y = a^{xy}$$

Raising a base number to a negative exponent is equivalent to 1 divided by the base number raised to the exponent's absolute value:

$$a^{-x} = \frac{1}{a^x}$$

To raise a base number to a fractional exponent, follow this formula:

$$a^{\frac{x}{y}} = \sqrt[y]{a^x}$$

Also keep in mind that any number other than 0 (zero) raised to the power of 0 (zero) equals 1:

$$a^0 = 1 \ [a \neq 0]$$

Example:

$(2^3)^2 \cdot 4^{-3} =$

(A) 16

(B) 1

(C) $\dfrac{2}{3}$

(D) $\dfrac{1}{2}$

(E) $\dfrac{1}{8}$

Solution:

The correct answer is (B). $(2^3)^2 \cdot 4^{-3} = 2^{(2)(3)} \cdot \dfrac{1}{4^3} = \dfrac{2^6}{4^3} = \dfrac{2^6}{2^6} = 1$

Exercise 4

Work out each problem. For questions 1–4, circle the letter that appears before your answer. Question 5 is a grid-in question.

1. $\dfrac{a^2 b}{b^2 c} \div \dfrac{a^2 c}{b c^2} =$

    (A) $\dfrac{1}{a}$

    (B) $\dfrac{1}{b}$

    (C) $\dfrac{b}{a}$

    (D) $\dfrac{c}{b}$

    (E) 1

2. $4^n + 4^n + 4^n + 4^n =$

    (A) 4^{4n}
    (B) 16^n
    (C) $4^{(n \cdot n \cdot n \cdot n)}$
    (D) $4^{(n+1)}$
    (E) 16^{4n}

3. Which of the following expressions is a simplified form of $(-2x^2)^4$?

    (A) $16x^8$
    (B) $8x^6$
    (C) $-8x^8$
    (D) $-16x^6$
    (E) $-16x^8$

4. If $x = -1$, then $x^{-3} + x^{-2} + x^2 + x^3 =$

    (A) -2
    (B) -1
    (C) 0
    (D) 1
    (E) 2

5. What integer is equal to $4^{3/2} + 4^{3/2}$?

5. FUNCTION NOTATION

In a *function* (or *functional relationship*), the value of one variable depends upon the value of, or is "a function of," another variable. In mathematics, the relationship can be expressed in various forms. The new SAT uses the form $y = f(x)$—where y is a function of x. (Specific variables used may differ.) To find the value of the function for any value x, substitute the x-value for x wherever it appears in the function.

Example:

If $f(x) = 2x - 6x$, then what is the value of $f(7)$?

Solution:

The correct answer is –28. First, you can combine $2x - 6x$, which equals $-4x$. Then substitute (7) for x in the function: $-4(7) = -28$. Thus, $f(7) = -28$.

A problem on the new SAT may ask you to find the value of a function for either a number value (such as 7, in which case the correct answer will also be a number value) or for a variable expression (such as $7x$, in which case the correct answer will also contain the variable x). A more complex function problem might require you to apply two different functions or to apply the same function twice, as in the next example.

Example:

If $f(x) = \dfrac{2}{x^2}$, then $f\left(\dfrac{1}{2}\right) \times f\left(\dfrac{1}{x}\right) =$

(A) $4x$

(B) $\dfrac{1}{8x}$

(C) $16x$

(D) $\dfrac{1}{4x^2}$

(E) $16x^2$

Solution:

The correct answer is (E). Apply the function to each of the two x-values (in the first instance, you'll obtain a numerical value, while in the second instance you'll obtain an variable expression):

$$f\left(\frac{1}{2}\right) = \frac{2}{\left(\frac{1}{2}\right)^2} = \frac{2}{\frac{1}{4}} = 2 \times 4 = 8$$

$$f\left(\frac{1}{x}\right) = \frac{2}{\left(\frac{1}{x}\right)^2} = \frac{2}{\left(\frac{1}{x^2}\right)} = 2x^2$$

Then, combine the two results according to the operation specified in the question:

$$f\left(\frac{1}{2}\right) \times f\left(\frac{1}{x}\right) = 8 \times 2x^2 = 16x^2$$

Exercise 5

Work out each problem. Circle the letter that appears before your answer.

1. If $f(x) = 2x\sqrt{x}$, then for which of the following values of x does $f(x) = x$?

    (A) $\dfrac{1}{4}$

    (B) $\dfrac{1}{2}$

    (C) 2

    (D) 4

    (E) 8

2. If $f(a) = a^{-3} - a^{-2}$, then $f\left(\frac{1}{3}\right) =$

    (A) $-\dfrac{1}{6}$

    (B) $\dfrac{1}{6}$

    (C) 6

    (D) 9

    (E) 18

3. If $f(x) = x^2 + 3x - 4$, then $f(2 + a) =$

    (A) $a^2 + 7a + 6$

    (B) $2a^2 - 7a - 12$

    (C) $a^2 + 12a + 3$

    (D) $6a^2 + 3a + 7$

    (E) $a^2 - a + 6$

4. If $f(x) = x^2$ and $g(x) = x + 3$, then $g(f(x)) =$

    (A) $x + 3$

    (B) $x^2 + 6$

    (C) $x + 9$

    (D) $x^2 + 3$

    (E) $x^3 + 3x^2$

5. If $f(x) = \frac{x}{2}$, then $f(x^2) \div \left(f(x)\right)^2 =$

    (A) x^3

    (B) 1

    (C) $2x^2$

    (D) 2

    (E) $2x$

6. FUNCTIONS—DOMAIN AND RANGE

A function consists of a *rule* along with two sets—called the *domain* and the *range*. The domain of a function $f(x)$ is the set of all values of x on which the function $f(x)$ is defined, while the range of $f(x)$ is the set of all values that result by applying the rule to all values in the domain.

By definition, a function must assign *exactly one* member of the range to each member of the domain, and must assign at least one member of the domain to each member of the range. Depending on the function's rule and its domain, the domain and range might each consist of a finite number of values; or either the domain or range (or both) might consist of an infinite number of values.

Example:

> In the function $f(x) = x + 1$, if the domain of x is the set $\{2,4,6\}$, then applying the rule that $f(x) = x + 1$ to all values in the domain yields the function's range: the set $\{3,5,7\}$. (All values other than 2, 4, and 6 are outside the domain of x, while all values other than 3, 5, and 7 are outside the function's range.)

Example:

> In the function $f(x) = x^2$, if the domain of x is the set of all real numbers, then applying the rule that $f(x) = x^2$ to all values in the domain yields the function's range: the set of all non-negative real numbers. (Any negative number would be outside the function's range.)

Exercise 6

Work out each problem. Circle the letter that appears before your answer.

1. If $f(x) = \sqrt{x+1}$, and if the domain of x is the set $\{3,8,15\}$, then which of the following sets indicates the range of $f(x)$?

    (A) $\{-4, -3, -2, 2, 3, 4\}$
    (B) $\{2, 3, 4\}$
    (C) $\{4, 9, 16\}$
    (D) $\{3, 8, 15\}$
    (E) $\{$all real numbers$\}$

2. If $f(a) = 6a - 4$, and if the domain of a consists of all real numbers defined by the inequality $-6 < a < 4$, then the range of $f(a)$ contains all of the following members EXCEPT:

    (A) -24
    (B) $\sqrt{\dfrac{1}{6}}$
    (C) 0
    (D) 4
    (E) 20

3. If the range of the function $f(x) = x^2 - 2x - 3$ is the set $R = \{0\}$, then which of the following sets indicates the largest possible domain of x ?

    (A) $\{-3\}$
    (B) $\{3\}$
    (C) $\{-1\}$
    (D) $\{3, -1\}$
    (E) all real numbers

4. If $f(x) = \sqrt{x^2 - 5x + 6}$, which of the following indicates the set of all values of x at which the function is NOT defined?

    (A) $\{x \mid x < 3\}$
    (B) $\{x \mid 2 < x < 3\}$
    (C) $\{x \mid x < -2\}$
    (D) $\{x \mid -3 < x < 2\}$
    (E) $\{x \mid x < -3\}$

5. If $f(x) = \sqrt[3]{\dfrac{1}{x}}$, then the largest possible domain of x is the set that includes

    (A) all non-zero integers.
    (B) all non-negative real numbers.
    (C) all real numbers except 0.
    (D) all positive real numbers.
    (E) all real numbers.

7. LINEAR FUNCTIONS—EQUATIONS AND GRAPHS

A *linear function* is a function f given by the general form $f(x) = mx + b$, in which m and b are constants. In algebraic functions, especially where defining a line on the xy-plane is involved, the variable y is often used to represent $f(x)$, and so the general form becomes $y = mx + b$. In this form, each x-value (member of the domain set) can be paired with its corresponding y-value (member of the range set) by application of the function.

Example:

In the function $y = 3x + 2$, if the domain of x is the set of all positive integers less than 5, then applying the function over the entire domain of x results in the following set of (x,y) pairs: $S = \{(1,5), (2,8), (3,11), (4,14)\}$.

In addition to questions requiring you to solve a system of linear equations (by using either the substitution or addition-subtraction method), the new SAT includes questions requiring you to recognize any of the following:

*   A linear function (equation) that defines two or more particular (x,y) pairs (members of the domain set and corresponding members of the range set). These questions sometimes involve real-life situations; you may be asked to construct a mathematical "model" that defines a relationship between, for example, the price of a product and the number of units of that product.

*   The graph of a particular linear function on the xy-plane

*   A linear function that defines a particular line on the xy-plane.

Variations on the latter two types of problems may involve determining the slope and/or y-intercept of a line defined by a function, or identifying a function that defines a given slope and y-intercept.

Example:

In the linear function f, if $f(-3) = 3$ and if the slope of the graph of f in the xy-plane is 3, what is the equation of the graph of f?
(A) $y = 3x - 3$
(B) $y = 3x + 12$
(C) $y = x - 6$
(D) $y = -x$
(E) $y = 3x - 12$

Solution:

The correct answer is (B). In the general equation $y = mx + b$, slope (m) is given as 3.   To determine b, substitute -3 for x and 3 for y, then solve for b: $3 = 3(-3) + b$; $12 = b$. Only in choice (B) does $m = 3$ and $b = 12$.

Exercise 7

Work out each problem. Circle the letter that appears before your answer.

1. XYZ Company pays its executives a starting salary of $80,000 per year. After every two years of employment, an XYZ executive receives a salary raise of $1,000. Which of the following equations best defines an XYZ executive's salary (S) as a function of the number of years of employment (N) at XYZ?

   (A) $S = \dfrac{1,000}{N} + 80,000$

   (B) $S = N + 80,000$

   (C) $S = \dfrac{80,000}{N} + 1,000$

   (D) $S = 1,000N + 80,000$

   (E) $S = 500N + 80,000$

2. In the linear function g, if $g(4) = -9$ and $g(-2) = 6$, what is the y-intercept of the graph of g in the xy-plane?

   (A) $-\dfrac{9}{2}$

   (B) $-\dfrac{5}{2}$

   (C) $\dfrac{2}{5}$

   (D) 1

   (E) $\dfrac{3}{2}$

3. If two linear function f and g have identical domains and ranges, which of the following, each considered individually, could describe the graphs of f and g in the xy-plane?

   I. two parallel lines
   II. two perpendicular lines
   III. two vertical lines
   (A) I only
   (B) I and II only
   (C) II only
   (D) II and III only
   (E) I, II, and III

4. In the xy-plane below, if the scales on both axes are the same, which of the following could be the equation of a function whose graph is l_1 ?

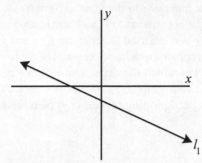

   (A) $y = \dfrac{2}{3}x - 3$

   (B) $y = -2x + 1$

   (C) $y = x + 3$

   (D) $y = -3x - \dfrac{2}{3}$

   (E) $y = -\dfrac{2}{3}x - 3$

5. If h is a linear function, and if $h(2) = 3$ and $h(4) = 1$, then $h(-101) =$

   (A) -72
   (B) -58
   (C) 49
   (D) 92
   (E) 106

8. QUADRATIC FUNCTIONS—EQUATIONS AND GRAPHS

In Chapter 8, you learned to solve quadratic equations in the general form $ax^2 + bx + c = 0$ by factoring the expression on the left-hand side of this equation to find the equation's two roots— the values of x that satisfy the equation. (Remember that the two roots might be the same.) The new SAT may also include questions involving *quadratic functions* in the general form $f(x) = ax^2 + bx + c$. (Note that a, b, and c are constants and that a is the only essential constant.) In quadratic functions, especially where defining a graph on the xy-plane is involved, the variable y is often used to represent $f(x)$, and x is often used to represent $f(y)$.

The graph of a quadratic equation of the basic form $y = ax^2$ or $x = ay^2$ is a *parabola*, which is a U-shaped curve. The point at which the dependent variable is at its minimum (or maximum) value is the *vertex*. In each of the following four graphs, the parabola's vertex lies at the origin (0,0). Notice that the graphs are constructed by tabulating and plotting several (x,y) pairs, and then connecting the points with a smooth curve:

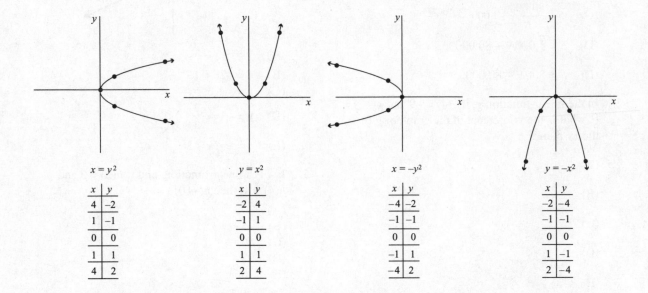

$x = y^2$	
x	y
4	−2
1	−1
0	0
1	1
4	2

$y = x^2$	
x	y
−2	4
−1	1
0	0
1	1
2	4

$x = -y^2$	
x	y
−4	−2
−1	−1
0	0
−1	1
−4	2

$y = -x^2$	
x	y
−2	−4
−1	−1
0	0
1	−1
2	−4

The graph of a quadratic equation of the basic form $x = \dfrac{1}{y^2}$ or $y = \dfrac{1}{x^2}$ is a *hyperbola*, which consists of two U-shaped curves that are symmetrical about a particular line, called the *axis of symmetry*. The axis of symmetry of the graph of $x = \dfrac{1}{y^2}$ is the x-axis, while the axis of symmetry in the graph of $y = \dfrac{1}{x^2}$ is the y-axis, as the next figure shows. Again, the graphs are constructed by tabulating and plotting some (x,y) pairs, then connecting the points:

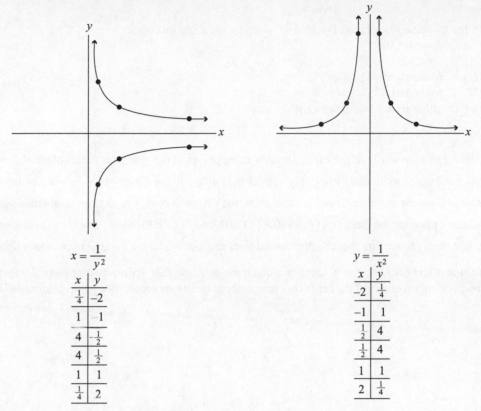

$$x = \frac{1}{y^2}$$

x	y
$\frac{1}{4}$	−2
1	−1
4	$-\frac{1}{2}$
4	$\frac{1}{2}$
1	1
$\frac{1}{4}$	2

$$y = \frac{1}{x^2}$$

x	y
−2	$\frac{1}{4}$
−1	1
$\frac{1}{2}$	4
$\frac{1}{2}$	4
1	1
2	$\frac{1}{4}$

The new SAT might include a variety of question types involving quadratic functions—for example, questions that ask you to recognize a quadratic equation that defines a particular graph in the xy-plane or to identify certain features of the graph of a quadratic equation, or compare two graphs

Example:

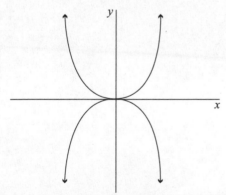

The graph shown in the xy-plane above could represent which of the following equations?

(A) $|x^2| = |y^2|$
(B) $x = |y^2|$
(C) $|y| = x^2$
(D) $y = |x^2|$
(E) $|x| = y^2$

Solution:

The correct answer is (C). The equation $|y| = x^2$ represents the union of the two equations $y = x^2$ and $-y = x^2$. The graph of $y = x^2$ is the parabola extending upward from the origin (0,0) in the figure, while the graph of $-y = x^2$ is the parabola extending downward from the origin.

Example:

In the xy-plane, the graph of $y + 2 = \dfrac{x^2}{2}$ shows a parabola that opens

(A)   downward.
(B)   upward.
(C)   to the right.
(D)   to the left.
(E)   either upward or downward.

Solution:

The correct answer is (B). Plotting three or more points of the graph on the xy-plane should show the parabola's orientation. First, it is helpful to isolate y in the equation $y = \dfrac{x^2}{2} - 2$. In this equation, substitute some simple values for x and solve for y in each case. For example, substituting 0, 2, and -2 for x gives us the three (x,y) pairs $(0,-2)$, $(2,0)$, and $(-2,0)$. Plotting these three points on the xy-plane, then connecting them with a curved line, suffices to show a parabola that opens upward.

An SAT question might also ask you to identify a quadratic equation that defines two or more domain members and the corresponding members of the function's range (these questions sometimes involve "models" of real-life situations).

Exercise 8

Work out each problem. Circle the letter that appears before your answer.

1. Which of the following equations defines a function containing the (x,y) pairs $(1,-1)$, $(2,-4)$, $(3,-9)$, and $(4,-16)$?

   (A) $y = -2x$
   (B) $y = 2x$
   (C) $y = x^2$
   (D) $y = -x^2$
   (E) $y = -2x^2$

2. The figure below shows a parabola in the xy-plane.

   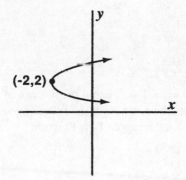

   Which of the following equations does the graph best represent?

   (A) $x = (y-2)^2 - 2$
   (B) $x = (y+2)^2 - 2$
   (C) $x = -(y-2)^2 - 2$
   (D) $y = (x-2)^2 + 2$
   (E) $y = (x-2)^2 - 2$

3. In the xy-plane, which of the following is an equation whose graph is the graph of $y = \dfrac{x^2}{3}$ translated three units horizontally and to the left?

   (A) $y = x^2$
   (B) $y = \dfrac{x^2}{3} + 3$
   (C) $y = \dfrac{x^2}{3} - 3$
   (D) $y = \dfrac{(x-3)^2}{3}$
   (E) $y = \dfrac{(x+3)^2}{3}$

4. Which of the following is the equations best defines the graph shown below in the xy-plane?

   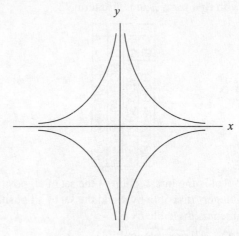

   (A) $y = \dfrac{1}{x^2}$

   (B) $x = \left| \dfrac{1}{y^2} \right|$

   (C) $x = \dfrac{1}{y^2}$

   (D) $|x| = \dfrac{1}{y^2}$

   (E) $y = \left| \dfrac{1}{x^2} \right|$

5. ABC Company projects that it will sell 48,000 units of product X per year at a unit price of $1, 12,000 units per year at $2 per unit, and 3,000 units per year at $4 per unit. Which of the following equations could define the projected number of units sold per year (N), as a function of price per unit (P)?

   (A) $N = \dfrac{48,000}{P^2 + 2}$

   (B) $N = \dfrac{48,000}{P^2}$

   (C) $N = \dfrac{48,000}{P + 14}$

   (D) $N = \dfrac{48,000}{P + 4}$

   (E) $N = \dfrac{48,000}{P^2 + 8}$

RETEST

Answer the following 10 questions, limiting your time to 15 minutes. Note that question 1 is a *grid-in* question, in which you provide the numerical solution. (All other questions are in multiple-choice format.)

1. What is the fourth term in a geometric series with first term 2 and third term 72?

    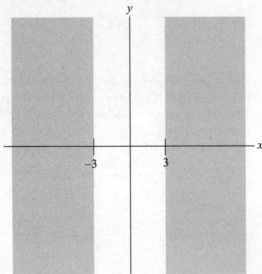

2. What is the intersection of the set of all positive integers divisible by 4 and the set of all positive integers divisible by 6?

    (A) All positive multiples of 4
    (B) All positive multiples of 6
    (C) All positive multiples of 8
    (D) All positive multiples of 12
    (E) All positive multiples of 24

3. The shaded regions of the *xy*-plane shown below represent certain values of *x*.

    Which of the following inequalities accounts for all such values of *x* ?

    (A) $|y| \geq 3$
    (B) $|x| \geq 3$
    (C) $|x| \leq 3$
    (D) $|y| \leq 3$
    (E) $|y| \leq -3$

4. If $-32 = \left(-\dfrac{1}{2}\right)^{M}$, then what is the value of *M* ?

    (A) −16
    (B) −6
    (C) −5
    (D) 5
    (E) 16

5. If $f(x) = \dfrac{1}{x+1}$, then $f\left(\dfrac{1}{x+1}\right) =$

    (A) x

    (B) $\dfrac{x+1}{2}$

    (C) 1

    (D) $x + 1$

    (E) $\dfrac{x+1}{x+2}$

6. If $f(x) = y = 1 - x^2$, and if the domain of *x* is all real numbers, which of the following sets indicates the range of the function?

    (A) $\{y \mid y \geq 1\}$
    (B) $\{y \mid y > 1\}$
    (C) $\{y \mid y \leq 1\}$
    (D) $\{y \mid y < 1\}$
    (E) $\{y \mid y \geq -2\}$

7. In the linear function *f*, if $f(-6) = -2$ and the slope of the graph of *f* in the *xy*-plane is −2, which of the following is true?

    (A) $f(-10) = -6$
    (B) $f(-6) = 0$
    (C) $f(-8) = 2$
    (D) $f(6) = 2$
    (E) $f(8) = 4$

8. Once a certain airplane attains its maximum speed of 300 miles per hour (mph), it begins decreasing speed as it approaches its destination. After every 50 miles, the plane decreases its airspeed by 20 mph. Which of the following equations best defines the number of miles the plane has traveled (m) after beginning to decrease speed as a function of the airplane's airspeed (s)?

(A) $s = -\dfrac{5m}{2} + 750$

(B) $s = -\dfrac{2m}{5} + 300$

(C) $m = -\dfrac{5s}{2} + 750$

(D) $m = -\dfrac{5s}{2} + 300$

(E) $m = \dfrac{2s}{5} + 300$

9. In the xy-plane, the graph of $3x = 2y^2$ shows a parabola with vertex at the point defined by the (x,y) pair:

(A) (0,0)
(B) (0,2)
(C) (2,0)
(D) (3,2)
(E) (2,3)

10. A model rocket is shot straight up in the air from ground level. After 2 seconds and then again after 3 seconds, its height is 96 feet. Which of the following equations could define rocket's height, (h), as a function of the number of seconds after launch (t)?

(A) $h = 10t^2 - 74t$
(B) $h = 8t^2 - 64t$
(C) $h = 64t - 8t^2$
(D) $h = 80t - 16t^2$
(E) $h = 96t - 10t^2$

SOLUTIONS TO PRACTICE EXERCISES

Diagnostic Test

1. The correct answer is 16,000. Solve for a in the general equation $a \cdot r^{(n-1)} = T$. Let $T = 256,000$, $r = 2$, and $n = 5$ (the number of terms in the sequence that includes the city's population 20, 15, 10, and 5 years ago, as well as its current population). Solving for a:

$$a \times 2^{(5-1)} = 256,000$$
$$a \times 2^4 = 256,000$$
$$a \times 16 = 256,000$$
$$a = 256,000 \div 16$$
$$a = 16,000$$

Twenty years ago, Urbanville's population was 16,000.

2. **(E)** The union of the three sets of factors is a set that contains all factors of any one or more of the three sets. The factors of 30 include all factors of 15, as well as the integer 6 and 30 (but not the integer 25). Choice (B) desribes the union of the factors of 15 and the factors of 30, but not the factors of 75 (which include 25). The factors of 75 include all factors of 15, as well as the integer 25 (but not integers 6 and 30). Choice (D) desribes the union of the factors of 15 and the factors of 75, but not the factors of 30 (which include 6 and 30). Thus, among answer choices (A) through (D), none describes the intersection of all three sets.

3. **(C)** First, determine each of the three absolute values:

$$|-1 - 2| = |-3| = 3$$
$$|5 - 6| = |-1| = 1$$
$$|-3 + 4| = |1| = 1$$

The combine the three results: $3 - 1 - 1 = 1$.

4. **(D)** Multiply like base numbers by adding exponents, and divide like base numbers by subtracting the denominator exponent from the numerator exponent:

$$\frac{x^6 y^3 y}{y^6 x^3 x} = \frac{x^6 y^4}{y^6 x^4} = \frac{x^2}{y^2}$$

5. **(B)** Substitute the expression $(x + 1)$ for x in $\frac{1}{f(x)} \times f\left(\frac{1}{x}\right)$, then combine terms:

$$\frac{1}{f(x)} \times f\left(\frac{1}{x}\right) = \left(\frac{1}{x+1}\right)\left(\frac{1+x}{x}\right) = \frac{1}{x}$$

6. **(D)** The question asks which number among the five listed is outside the function's range. First, simplify the function. Note that $5^{-x} = \frac{1}{5^x}$, and therefore that $\frac{x}{5^{-x}} = \frac{x}{\frac{1}{5^x}} = (x)(5^x)$. To determine the function's range, apply the rule that $f(x) = (x)(5^x)$ to each member of the domain, in turn:

$$f(-2) = (-2)(5^{-2}) = \frac{-2}{5^2} = -\frac{2}{25}$$
$$f(-1) = (-1)(5^{-1}) = \frac{-1}{5^1} = -\frac{1}{5}$$
$$f(0) = (0)(5^0) = (0)(1) = 0$$
$$f(2) = (2)(5^2) = (2)(25) = 50$$

The range of $f(x) = \frac{x}{5^{-x}}$ is the set $\{-\frac{2}{25}, -\frac{1}{5}, 0, 50\}$. Only answer choice (D) provides a number that is not in this range.

7. **(B)** To solve this problem, consider each answer choice in turn, substituting the (x,y) pairs provided in the question for x and y in the equation. Among the five equations, only the equation in choice (B) holds for *both* pairs:

$$y = 2x + 1$$
$$(-1) = 2(-1) + 1$$
$$(0) = 2(-\tfrac{1}{2}) + 1$$

8. **(A)** One point on l is defined by $(-5, -6)$. A second point on l is defined by $(4, 0)$, which is the point of x-intercept. With two points defined, you can find the line's slope (m) as follows:

$$m = \frac{y_2 - y_1}{x_2 - x_1} = \frac{0 - (-6)}{4 - (-5)} = \frac{6}{9} = \frac{2}{3}.$$

9. **(C)** The graph shows a parabola opening upward with vertex at (3,0). Of the five choices, only (A) and (C) provide equations that hold for the (x,y) pair (3,0). Eliminate choices (B), (D), and (E). In the equation given by choice (A), substituting any non-zero number for x yields a *negative* y-value. However, the graph shows no negative y-values. Thus, you can eliminate choice (A), and the correct answer must be (C).

Also, when a parabola extends upward, the coefficient of x^2 in the equation must be positive.

10. **(E)** The following figure shows the graphs of the two equations:

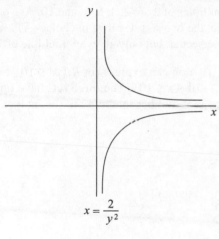

$$x = \frac{2}{y^2}$$

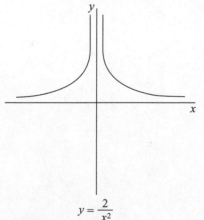

$$y = \frac{2}{x^2}$$

As you can see, the graphs are not mirror images of each other about any of the axes described in answer choices (A) through (D).

Exercise 1

1. **(E).** Solve for T in the general equation $a \cdot r^{(n-1)} = T$. Let $a = 1{,}500$, $r = 2$, and $n = 6$ (the number of terms in the sequence that includes the value in 1950 and at every 12-year interval since then, up to and including the expected value in 2010). Solving for T:

$$1{,}500 \times 2^{(6-1)} = T$$
$$1{,}500 \times 2^5 = T$$
$$1{,}500 \times 32 = T$$
$$48{,}000 = T$$

Doubling every 12 years, the land's value will be $48,000 in 2010.

2. **(A)** Solve for T in the general equation $a \cdot r^{(n-1)} = T$. Let $a = 4$, $r = 2$, and $n = 9$ (the number of terms in the sequence that includes the number of cells observable now as well as in 4, 8, 12, 16, 20, 24, 28, and 32 seconds). Solving for T:

$$4 \times 2^{(9-1)} = T$$
$$4 \times 2^8 = T$$
$$4 \times 256 = T$$
$$1{,}024 = T$$

32 seconds from now, the number of observable cancer cells is 1,024.

3. **(B)** In the standard equation, let $T = 448$, $r = 2$, and $n = 7$. Solve for a:

$$a \times 2^{(7-1)} = 448$$
$$a \times 2^6 = 448$$
$$a \times 64 = 448$$
$$a = \frac{448}{64}$$
$$a = 7$$

4. The correct answer is $800. Solve for a in the general equation $a \cdot r^{(n-1)} = T$. Let $T = 2,700$. The value at the date of the purchase is the first term in the sequence, and so the value three years later is the fourth term; accordingly, $n = 4$. Given that painting's value increased by 50% (or $\frac{1}{2}$) per year on average, $r = 1.5 = \frac{3}{2}$. Solving for a:

$$a \times \left(\frac{3}{2}\right)^{(4-1)} = 2,700$$

$$a \times \left(\frac{3}{2}\right)^{3} = 2,700$$

$$a \times \frac{27}{8} = 2,700$$

$$a = 2,700 \times \frac{8}{27}$$

$$a = 800$$

At an increase of 50% per year, the collector must have paid $800 for the painting three years ago.

5. The correct answer is 21. First, find r:

$$3 \times r^{(3-1)} = 147$$

$$3 \times r^{2} = 147$$

$$r^{2} = 49$$

$$r = 7$$

To find the second term in the sequence, multiply the first term (3) by r: $3 \cdot 7 = 21$.

Exercise 2

1. **(E)** The union of the two sets is the set that contains all integers — negative, positive, and zero (0).

2. The correct answer is 4. The positive factors of 24 are 1, 2, 3, 4, 6, 8, 12, and 24. The positive factors of 18 are 1, 2, 3, 6, 9, and 18. The two sets have in common four members: 1, 2, 3, and 6.

3. **(C)** 19 is a prime number, and therefore has only one prime factor: 19. There are two prime factors of 38: 2 and 19. The union of the sets described in choice (C) is the set that contains two members: 2 and 19.

4. **(A)** Through 10, the multiples of $\frac{5}{2}$, or $2\frac{1}{2}$, are $2\frac{1}{2}$, 5, $2\frac{1}{2}$, and 10. Through 10, the multiples of 2 are 2, 4, 6, 8, and 10. As you can see, the two sets desribed in choice (A) intersect at, but only at, every multiple of 10.

5. **(D)** You can express set $R:\{|x| \leq 10\}$ as $R:\{-10 \leq x \leq 10\}$. The three sets have only one real number in common: the integer 10.

Exercise 3

1. **(D)** $|7 - 2| - |2 - 7| = |5| - |-5| = 5 - 5 = 0$

2. **(C)** If $b - a$ is a negative integer, then $a > b$, in which case $a - b$ must be a positive integer. (When you subtract one integer from another, the result is always an integer.) Choice (A), which incorporates the concept of absolute value, cannot be the correct answer, since the absolute value of any integer is by definition a positive integer.

3. **(E)** Either $x - 3 > 4$ or $x - 3 < -4$. Solve for x in both inequalities: $x > 7$; $x < -1$.

4. **(B)** If $x = 0$, $y = -1$. The point $(0,-1)$ on the graph shows this functional pair. For all negative values of x, y is the absolute value of x, minus 1 (the graph is translated down one unit). The portion of the graph to the left of the y-axis could show these values. For all positive values of x, $y = x$, minus 1 (the graph is translated down one unit). The portion of the graph to the right of the y-axis could show these values.

5. **(D)** Substitute $\frac{1}{2}$ for x in the function:

$$f\left(\tfrac{1}{2}\right) = \left|\frac{\frac{1}{2}}{\frac{1}{2}} - 3\right| - \frac{1}{2}$$
$$= |2 - 3| - \frac{1}{2}$$
$$= |-1| - \frac{1}{2}$$
$$= 1 - \frac{1}{2}$$
$$= \frac{1}{2}$$

Exercise 4

1. **(E)** First, cancel common factors in each term. Then, multiply the first term by the reciprocal of the second term. You can now see that all terms cancel out:

$$\frac{a^2 b}{b^2 c} \div \frac{a^2 c}{bc^2} = \frac{a^2}{bc} \div \frac{a^2}{bc} = \frac{a^2}{bc} \times \frac{bc}{a^2} = 1$$

2. **(D)** The expression given in the question is equivalent to $4 \cdot 4^n$. In this expression, base numbers are the same. Since the terms are multiplied together, you can combine exponents by adding them together: $4 \cdot 4^n = 4^{(n+1)}$.

3. **(A)** Raise both the coefficient -2 and variable x^2 to the power of 4. When raising an exponent to a power, multiply together the exponents:

$$(-2x^2)^4 = (-2)^4 x^{(2)(4)} = 16x^8$$

4. **(C)** Any term to a negative power is the same as "one over" the term, but raised to the *positive* power. Also, a negative number raised to a power is negative if the exponent is *odd*, yet positive if the exponent is *even*:

$$-1^{(-3)} + [-1^{(-2)}] + [-1^2] + [-1^3] = -\frac{1}{1} + \frac{1}{1} + 1 - 1$$
$$= 0$$

5. The correct answer is 16. Express fractional exponents as roots, calculate the value of each term, and then add:

$$4^{3/2} + 4^{3/2} = \sqrt{4^3} + \sqrt{4^3} = \sqrt{64} + \sqrt{64} = 8 + 8 = 16$$

Exercise 5

1. **(A)** One way to approach this problem is to substitute each answer choice for x in the function, then find $f(x)$. Only choice (A) provides a value for which $f(x) = x$:

    $$f\left(\tfrac{1}{4}\right) = 2\left(\tfrac{1}{4}\right)\sqrt{\left(\tfrac{1}{4}\right)} = \left(\tfrac{1}{2}\right)\left(\tfrac{1}{2}\right) = \tfrac{1}{4}$$

    Another way to solve the problem is to let $x = 2x\sqrt{x}$, then solve for x by squaring both sides of the equation (look for a root that matches one of the answer choices):

    $$x = 2x\sqrt{x}$$
    $$1 = 2\sqrt{x}$$
    $$\frac{1}{2} = \sqrt{x}$$
    $$\frac{1}{4} = x$$

2. **(E)** First, note that any term raised to a negative power is equal to 1 divided by the term to the absolute value of the power. Hence:

    $$a^{-3} - a^{-2} = \frac{1}{a^3} - \frac{1}{a^2}$$

    Using this form of the function, substitute $\frac{1}{3}$ for a, then simplify and combine terms:

    $$f\left(\tfrac{1}{3}\right) = \frac{1}{\left(\tfrac{1}{3}\right)^3} - \frac{1}{\left(\tfrac{1}{3}\right)^2} = \frac{1}{\tfrac{1}{27}} - \frac{1}{\tfrac{1}{9}} = 27 - 9 = 18$$

3. **(A)** In the function, substitute $(2 + a)$ for x. Since each of the answer choices indicates a quadratic expression, apply the distributive property of arithmetic, then combine terms:

    $$f(2+a) = (2+a)^2 + 3(2+a) - 4$$
    $$= (2+a)(2+a) + 6 + 3a - 4$$
    $$= 4 + 4a + a^2 + 6 + 3a - 4$$
    $$= a^2 + 7a + 6$$

4. **(D)** Substitute $f(x)$ for x in the function $g(x) = x + 3$:

    $$g(f(x)) = f(x) + 3$$

    Then substitute x^2 for $f(x)$:

    $$g(f(x)) = x^2 + 3$$

5. **(D)** $f(x^2) = \frac{x^2}{2}$, and $\left(f(x)\right)^2 = \left(\frac{x}{2}\right)^2$.

    Accordingly, $f(x^2) \div \left(f(x)\right)^2 = \frac{x^2}{2} \div \left(\frac{x}{2}\right)^2 = \frac{x^2}{2} \cdot \frac{4}{x^2} = 2$.

Exercise 6

1. **(B)** To determine the function's range, apply the rule $\sqrt{x+1}$ to 3, 8, and 15:

    $$\sqrt{(3)+1} = \sqrt{4} = +2$$
    $$\sqrt{(8)+1} = \sqrt{9} = +3$$
    $$\sqrt{(15)+1} = \sqrt{16} = +4$$

    Choice (B) provides the members of the range. Remember that $\sqrt{x}$ means the *positive* square root of x.

2. **(E)** To determine the function's range, apply the rule $(6a - 4)$ to –6 and to 4. The range consists of all real numbers between the two results:

    $$6(-6) - 4 = -40$$
    $$6(4) - 4 = 20$$

    The range of the function can be expressed as the set $R = \{b \mid -40 < b < 20\}$. Of the five answer choices, only (E) does not fall within the range.

3. **(D)** The function's range contains only one member: the number 0 (zero). Accordingly, to find the domain of x, let $f(x) = 0$, and solve for all possible roots of x:

    $$x^2 - 2x - 3 = 0$$
    $$(x-3)(x+1) = 0$$
    $$x - 3 = 0, \; x + 1 = 0$$
    $$x = 3, \; x = -1$$

    Given that $f(x) = 0$, the largest possible domain of x is the set $\{3, -1\}$.

4. **(B)** The question asks you to recognize the set of values outside the domain of x. To do so, first factor the trinomial within the radical into two binomials:

    $$f(x) = \sqrt{x^2 - 5x + 6} = \sqrt{(x-3)(x-2)}$$

    The function is undefined for all values of x such that $(x - 3)(x - 2) < 0$ because the value of the function would be the square root of a negative number (not a real number). If $(x - 3)(x - 2) < 0$, then one binomial value must be negative while the other is positive. You obtain this result with any value of x greater than 2 but less than 3—that is, when $2 < x < 3$.

5. **(C)** If $x = 0$, then the value of the fraction is undefined; thus, 0 is outside the domain of x. However, the function can be defined for any other real-number value of x. (If $x > 0$, then applying the function yields a positive number; if $x < 0$, then applying the function yields a negative number.)

Exercise 7

1. **(E)** After the first 2 years, an executive's salary is raised from \$80,000 to \$81,000. After a total of 4 years, that salary is raised to \$82,000. Hence, two of the function's (N,S) pairs are (2, \$81,000) and (4, \$82,000). Plugging both of these (N,S) pairs into each of the five equations, you see that only the equation in choice (E) holds (try plugging in additional pairs to confirm this result):

   $(81,000) = (500)(2) + 80,000$

   $(82,000) = (500)(4) + 80,000$

   $(83,000) = (500)(6) + 80,000$

2. **(D)** The points (4,–9) and (–2,6) both lie on the graph of g, which is a straight line. The question asks for the line's y-intercept (the value of b in the general equation $y = mx + b$). First, determine the line's slope:

   $$\text{slope } (m) = \frac{y_2 - y_1}{x_2 - x_1} = \frac{6 - (-9)}{-2 - 4} = \frac{15}{-6} = -\frac{5}{2}$$

   In the general equation $(y = mx + b)$, $m = -\frac{5}{2}$. To find the value of b, substitute either (x,y) value pair for x and y, then solve for b. Substituting the (x,y) pair (–2,6):

   $y = -\frac{5}{2}x + b$

   $6 = -\frac{5}{2}(-2) + b$

   $6 = 5 + b$

   $1 = b$

3. **(B)** In the xy-plane, the domain and range of any line other than a vertical or horizontal line is the set of all real numbers. Thus, option III (two vertical lines) is the only one of the three options that *cannot* describe the graphs of the two functions.

4. **(E)** The line shows a negative y-intercept (the point where the line crosses the vertical axis) and a negative slope less than –1 (that is, slightly more horizontal than a 45° angle). In equation (E), $-\frac{2}{3}$ is the slope and –3 is the y-intercept. Thus, equation (E) matches the graph of the function.

5. **(E)** The function h includes the two functional pairs $(2,3)$ and $(4,1)$. Since h is a linear function, its graph on the xy-plane is a straight line. You can determine the equation of the graph by first finding its slope (m):

$$m = \frac{y_2 - y_1}{x_2 - x_1} = \frac{1-3}{4-2} = \frac{-2}{2} = -1 .$$

Plug either (x,y) pair into the standard equation $y = mx + b$ to define the equation of the line. Using the pair $(2,3)$:

$$y = -x + b$$
$$3 = -2 + b$$
$$5 = b$$

The line's equation is $y = -x + 5$. To determine which of the five answer choices provides a point that also lies on this line, plug in the value -101 (as provided in the question) for x:

$$y = -(-101) + 5 = 101 + 5 = 106.$$

Exercise 8

1. **(D)** To solve this problem, consider each answer choice in turn, substituting the (x,y) pairs provided in the question for x and y in the equation. Among the five equations, only the equation in choice (D) holds for all four pairs.

2. **(A)** The graph shows a parabola opening to the right with vertex at $(-2,2)$. If the vertex were at the origin, the equation defining the parabola might be $x = y^2$. Choices (D) and (E) define vertically oriented parabolas (in the general form $y = x^2$) and thus can be eliminated. Considering the three remaining equations, (A) and (C) both hold for the (x,y) pair $(-2,2)$, but (B) does not. Eliminate (B). Try substituting 0 for y in equations (A) and (C), and you'll see that only in equation (A) is the corresponding x-value greater than 0, as the graph suggests.

3. **(E)** The equation $y = \frac{x^2}{3}$ is a parabola with vertex at the origin and opening upward. To see that this is the case, substitute some simple values for x and solve for y in each case. For example, substituting 0, 3, and -3 for x gives us the three (x,y) pairs $(0,0)$, $(3,3)$, and $(-3,3)$. Plotting these three points on the xy-plane, then connecting them with a curved line, suffices to show a parabola with vertex $(0,0)$ — opening upward. Choice (E) provides an equation whose graph is identical to the graph of $y = \frac{x^2}{3}$, except translated three units to the left. To confirm this, again, substitute simple values for x and solve for y in each case. For example, substituting 0, -3, and -6 for x gives us the three (x,y) pairs $(0,3)$, $(-3,0)$, and $(-6,-3)$. Plotting these three points on the xy-plane, then connecting them with a curved line, suffices to show the same parabola as the previous one, except with vertex $(-3,0)$ instead of $(0,0)$.

4. **(D)** The equation $|x| = \frac{1}{y^2}$ represents the union of the two equations $x = \frac{1}{y^2}$ and $-x = \frac{1}{y^2}$. The graph of the former equation is the hyperbola shown to the right of the y-axis in the figure, while the graph of the latter equation is the hyperbola shown to the left of the y-axis in the figure.

5. **(B)** In this problem, S is a function of P. The problem provides three (P,S) number pairs that satisfy the function: $(1, 48{,}000)$, $(2, 12{,}000)$ and $(4, 3{,}000)$. For each of the answer choices, plug each of these three (P,S) pairs in the equation given. Only the equation given in choice (B) holds for all three (P,S) pairs:

$$48{,}000 = \frac{48{,}000}{(1)^2} = 48{,}000$$

$$12{,}000 = \frac{48{,}000}{(2)^2} = \frac{48{,}000}{4} = 12{,}000$$

$$3{,}000 = \frac{48{,}000}{(4)^2} = \frac{48{,}000}{16} = 3{,}000$$

Retest

1. The correct answer is 432. First, find r:

$$2 \times r^{(3-1)} = 72$$
$$2 \times r^2 = 72$$
$$r^2 = 36$$
$$r = 6$$

To find the fourth term in the sequence, solve for T in the standard equation (let $r = 6$ and $n = 4$):

$$2 \times 6^{(4-1)} = T$$
$$2 \times 6^3 = T$$
$$2 \times 216 = T$$
$$432 = T$$

2. **(D)** The set of positive integers divisible by 4 includes all multiples of 4: 4, 8, 12, 16, ... The set of positive integers divisible by 6 includes all multiples of 6: 6, 12, 18, 24, ... The least common multiple of 4 and 6 is 12. Thus, common to the two sets are all multiples of 12, but no other elements.

3. **(B)** The shaded region to the left of the y-axis accounts for all values of x that are less than or equal to -3. In other words, this region is the graph of $x \leq -3$. The shaded region to the right of the y-axis accounts for all values of x that are greater than or equal to 3. In other words, this region is the graph of $x \geq 3$.

4. **(C)** Note that $(-2)^5 = -32$. So, the answer to the problem must involve the number 5. However, the 2 in the number $\frac{1}{2}$ is in the denominator, and you must move it to the numerator. Since a negative number reciprocates its base, $\left(-\frac{1}{2}\right)^{-5} = -32$.

5. **(E)** Substitute $\frac{1}{x+1}$ for x, then simplify:

$$f\left(\frac{1}{x+1}\right) = \frac{1}{\frac{1}{x+1}+1} = \frac{1}{\frac{1}{x+1}+\frac{x+1}{x+1}} = \frac{1}{\frac{1+(x+1)}{x+1}}$$
$$= \frac{x+1}{1+(x+1)} = \frac{x+1}{x+2}$$

6. **(C)** According to the function, if $x = 0$, then $y = 1$. (The function's range includes the number 1.) If you square any real number x other than 0, the result is a number greater than 0. Accordingly, for any non-zero value of x, $1 - x^2 < 1$. The range of the function includes 1 and all numbers less than 1.

7. **(C)** The graph of f is a straight line, one point on which is $(-6,-2)$. In the general equation $y = mx + b$, $m = -2$. To find the value of b, substitute the (x,y) value pair $(-6,-2)$ for x and y, then solve for b:

$$y = -2x + b$$
$$(-2) = -2(-6) + b$$
$$-2 = 12 + b$$
$$-14 = b$$

The equation of the function's graph is $y = -2x - 14$. Plugging in each of the five (x,y) pairs given, you can see that this equation holds only for choice (C).

8. **(C)** You can easily eliminate choices (A) and (B) because each one expresses speed (s) as a function of miles (m), just the reverse of what the question asks for. After the first 50 miles, the plane's speed decreases from 300 mph to 280 mph. After a total of 100 miles, the speed has decreased to 260. Hence, two of the function's (s,m) pairs are $(280,50)$ and $(260,100)$. Plugging both of these (s,m) pairs into each of the five equations, you see that only the equation in choice (C) holds (try plugging in additional pairs to confirm this result):

$$(50) = -\frac{5(280)}{2} + 750$$
$$50 = -\frac{1400}{2} + 750$$
$$50 = -700 + 750$$
$$50 = 50$$

$$(100) = -\frac{5(260)}{2} + 750$$
$$100 = --\frac{1300}{2} + 750$$
$$100 = -650 + 750$$
$$100 = 100$$

9. **(A)** The graph of any quadratic equation of the incomplete form $x = ay^2$ (or $y = ax^2$) is a parabola with vertex at the origin $(0,0)$. Isolating x in the equation $3x = 2y^2$ shows that the equation is of that form:

$$x = \frac{2y^2}{3}$$

To confirm that the vertex of the graph of $x = \frac{2y^2}{3}$ lies at $(0,0)$, substitute some simple values for y and solve for x in each case. For example, substituting 0, 1, and –1 for y gives us the three (x,y) pairs $(0,0)$, $(\frac{2}{3},1)$, and $(\frac{2}{3},-1)$. Plotting these three points on the xy-plane, then connecting them with a curved line, suffices to show a parabola with vertex $(0,0)$ — opening to the right.

10. **(D)** The question provides two (t,h) number pairs that satisfy the function: $(2,96)$ and $(3,96)$. For each of the answer choices, plug each of these two (t,h) pairs in the equation given. Only the equation given in choice (D) holds for both (t,h) pairs:

$$(96) = 80(2) - 16(2)^2 = 160 - 64 = 96$$

$$(96) = 80(3) - 16(3)^2 = 240 - 144 = 96$$

Note that the equation in choice (C) holds for $f(2) = 96$ but *not* for $f(3) = 96$.

DIAGNOSTIC TEST

Directions: Answer multiple-choice questions 1–11, as well as question 12, which is a "grid-in" (student-produced response) question. Try to answer questions 1 and 2 using trigonometry.

Answers are at the end of the chapter.

1. In the triangle shown below, what is the value of x?

    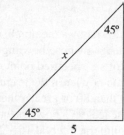

    (A) $4\sqrt{3}$
    (B) $5\sqrt{2}$
    (C) 8
    (D) $6\sqrt{2}$
    (E) $5\sqrt{3}$

2. In the triangle shown below, what is the value of x?

    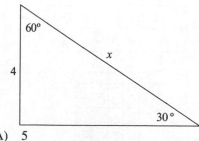

    (A) 5
    (B) 6
    (C) $4\sqrt{3}$
    (D) 8
    (E) $6\sqrt{2}$

3. The figure below shows a regular hexagon tangent to circle O at six points.

    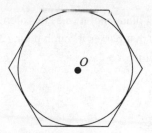

    If the area of the hexagon is $6\sqrt{3}$, the circumference of circle $O =$

    (A) $\dfrac{3\pi\sqrt{3}}{2}$

    (B) $\dfrac{12\sqrt{3}}{\pi}$

    (C) $2\pi\sqrt{3}$

    (D) 12

    (E) 6π

4. In the *xy*-plane, which of the following (*x*,*y*) pairs defines a point that lies on the same line as the two points defined by the pairs (2,3) and (4,1)?

   (A) (7,–3)
   (B) (–1,8)
   (C) (–3,2)
   (D) (–2,–4)
   (E) (6,–1)

5. In the *xy*-plane, what is the slope of a line that is perpendicular to the line segment connecting points A(–4,–3) and B(4,3)?

   (A) $-\dfrac{3}{2}$

   (B) $-\dfrac{4}{3}$

   (C) 0

   (D) $\dfrac{3}{4}$

   (E) 1

6. In the *xy*-plane, point (*a*,5) lies along a line of slope $\dfrac{1}{3}$ that passes through point (2,–3). What is the value of *a* ?

   (A) –26
   (B) –3
   (C) 3
   (D) 26
   (E) 35

7. The figure below shows the graph of a certain equation in the *xy*-plane. At how many different values of *x* does *y* = 2 ?

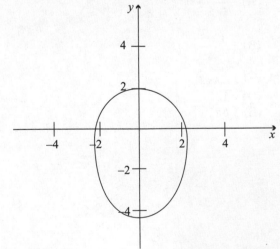

   (A) 0
   (B) 1
   (C) 2
   (D) 4
   (E) Infinitely many

8. If *f*(*x*) = *x*, then the line shown in the *xy*-plane below is the graph of

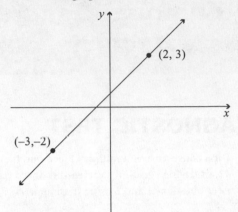

   (A) *f*(–*x*)
   (B) *f*(*x* + 1)
   (C) *f*(*x* – 1)
   (D) *f*(1 – *x*)
   (E) *f*(–*x* – 1)

9. The table below shows the number of bowlers in a certain league whose bowling averages are within each of six specified point ranges, or intervals. If no bowler in the league has an average less than 80 or greater than 200, what percent of the league's bowlers have bowling averages within the interval 161–200?

Interval	Frequency
80–100	17
101–120	56
121–140	48
141–160	59
161–180	37
181–200	23

10. Average annual rainfall and temperatures for five cities are plotted in the figure below. The cities are labeled by the letters A through E in order according to their east-west location; for example, City A is further east than City B, which is further east than City C. Based on the figure, which of the following statements is most accurate?

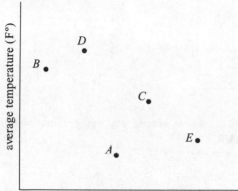

(A) The further west a city, the more annual rainfall it receives.

(B) The further east a city, the higher its average annual temperature.

(C) The more annual rainfall a city receives, the lower its average annual temperature.

(D) The higher a city's average annual temperature, the more annual rainfall it receives.

(E) The further east a city, the lower its average annual temperature.

11. One marble is to be drawn randomly from a bag that contains three red marbles, two blue marbles, and one green marble. What is the probability of drawing a blue marble?

(A) $\frac{1}{6}$

(B) $\frac{1}{5}$

(C) $\frac{2}{7}$

(D) $\frac{1}{3}$

(E) $\frac{2}{5}$

12. The figure below shows two concentric circles, each divided into eight congruent segments. The area of the large circle is exactly twice that of the smaller circle. If a point is selected at random from the large circular region, what is the probability that the point will lie in a shaded portion of that circle?

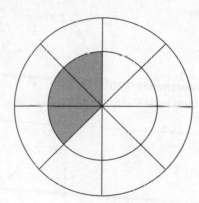

1. RIGHT TRIANGLES AND TRIGONOMETRIC FUNCTIONS

Right-triangle trigonometry involves the ratios between sides of right triangles and the angle measures that correspond to these ratios. Refer to the following right triangle, in which the sides opposite angles A, B, and C are labeled a, b, and c, respectively ($\angle A$ and $\angle B$ are the two acute angles):

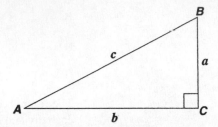

Here are the general definitions of the three trigonometric functions *sine*, *cosine*, and *tangent*, and how you would express these three functions in terms of $\angle A$ and $\angle B$ in $\triangle ABC$:

$$\text{sine} = \frac{\text{opposite}}{\text{hypotenuse}} \quad (\sin A = \frac{a}{c} \; ; \sin B = \frac{b}{c})$$

$$\text{cosine} = \frac{\text{adjacent}}{\text{hypotenuse}} \quad (\cos A = \frac{b}{c} \; ; \cos B = \frac{a}{c})$$

$$\text{tangent} = \frac{\text{opposite}}{\text{adjacent}} \quad (\tan A = \frac{a}{b} \; ; \tan B = \frac{b}{a})$$

In right triangles with angles 45°-45°-90° and 30°-60°-90°, the values of these trigonometric functions are easily determined. The following figure shows the ratios among the sides of these two uniquely shaped triangles:

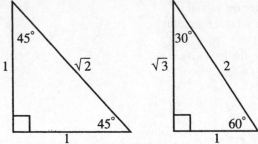

In a 45°-45°-90° triangle, the lengths of the sides opposite those angles are in the ratio $1:1:\sqrt{2}$, respectively. In a 30°-60°-90° triangle, the lengths of the sides opposite those angles are in the ratio $1:\sqrt{3}:2$, respectively. Accordingly, the sine, cosine, and tangent functions of the 30°, 45°, and 60° angles of *any right* triangle are as follows:

45°-45°-90° triangle:	**30°-60°-90° triangle:**
$\sin 45° = \cos 45° = \dfrac{\sqrt{2}}{2}$	$\sin 30° = \cos 60° = \dfrac{1}{2}$
$\tan 45° = 1$	$\sin 60° = \cos 30° = \dfrac{\sqrt{3}}{2}$
	$\tan 30° = \dfrac{\sqrt{3}}{3}$
	$\tan 60° = \sqrt{3}$

In SAT problems involving 30°-60°-90° and 45°-45°-90° right triangles, as long as the length of one side is provided, you can use these trigonometric functions to determine the length of any other side—as an alternative to applying the Pythagorean Theorem.

Example:

In the triangle shown below, what is the value of x ?

(A) $\sqrt{3}$

(B)   2

(C) $\dfrac{3\sqrt{2}}{2}$

(D) $\dfrac{5\sqrt{2}}{3}$

(E) $2\sqrt{2}$

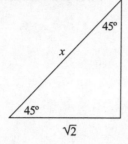

Solution:

The correct answer is (B). Since the figure shows a 45°-45°-90° triangle in which the length of one leg is known, you can easily apply either the sine or cosine function to determine the length of the hypotenuse. Applying the function $\sin 45° = \dfrac{\sqrt{2}}{2}$, set the value of this function equal to $\dfrac{\sqrt{2}}{x}\left(\dfrac{\text{opposite}}{\text{hypotenuse}}\right)$, then solve for x: $\dfrac{\sqrt{2}}{2} = \dfrac{\sqrt{2}}{x}$; $\sqrt{2}x = 2\sqrt{2}$; $x = 2$.

Example:

In the triangle shown below, what is the value of x ?

(A) $\dfrac{5\sqrt{3}}{3}$

(B)   3

(C) $\dfrac{10}{3}$

(D) $2\sqrt{3}$

(E) $3\sqrt{2}$

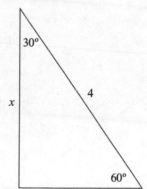

Solution:

The correct answer is (D). Since the figure shows a 30°-60°-90° triangle, you can easily apply either the sine or the cosine function to determine the length of either leg. Applying the function $\sin 60° = \dfrac{\sqrt{3}}{2}$, set the value of this function equal to $\dfrac{x}{4}\left(\dfrac{\text{opposite}}{\text{hypotenuse}}\right)$, then solve for x:

$$\frac{\sqrt{3}}{2} = \frac{x}{4} \; ; \; 2x = 4\sqrt{3} \; ; \; x = 2\sqrt{3}$$

Exercise 1

Work out each problem. Circle the letter that appears before your answer.

1. In the triangle shown below what is the value of *x* ?

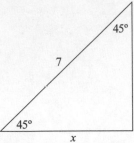

   (A) $3\sqrt{2}$

   (B) $\dfrac{7\sqrt{2}}{2}$

   (C) 5

   (D) $3\sqrt{3}$

   (E) $4\sqrt{2}$

2. In the triangle shown below, what is the value of *x* ?

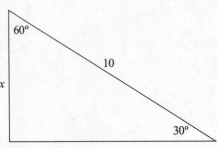

   (A) 4

   (B) $3\sqrt{2}$

   (C) 5

   (D) $3\sqrt{3}$

   (E) $\dfrac{10}{\sqrt{3}}$

3. If two interior angles of a triangle measure 30° and 60°, and if the side of the triangle opposite the 60° angle is 6 units long, how many units long is the side opposite the 30° angle?

   (A) 3

   (B) $2\sqrt{3}$

   (C) 4

   (D) $3\sqrt{2}$

   (E) $\dfrac{5\sqrt{3}}{2}$

4. What is the area of the triangle shown below?

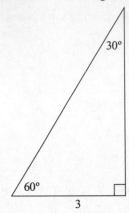

   (A) $5\sqrt{2}$

   (B) 7.5

   (C) $\dfrac{13\sqrt{3}}{3}$

   (D) $\dfrac{9\sqrt{3}}{2}$

   (E) $5\sqrt{3}$

5. Two trains depart at the same time from the same terminal, one traveling due north and the other due east, each along a straight track. If the trains travel at the same average speed, which of the following most closely approximates the number of miles each train has traveled when the shortest distance between the two trains is 70 miles?

   (A) 49

   (B) 55

   (C) 60

   (D) 70

   (E) 140

2. TANGENT LINES AND INSCRIBED CIRCLES

A circle is *tangent* to a line (or line segment) if it intersects the line (or line segment) at one and only one point (called the *point of tangency*). In addition to the rules you learned in Chapter 13 involving tangents, for the new SAT you should know the following two rules:

1. A line (or line segment) that is tangent to a circle is *always* perpendicular to a radius drawn from the circle's center to the point of tangency. Thus, in the next figure, which shows a circle with center O, $\overline{OP} \perp \overline{AB}$:

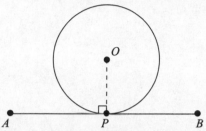

2. For any *regular* polygon (in which all sides are congruent) that circumscribes a circle, the point of tangency between each line segment and the circle *bisects* the segment. Thus, in the next figure, which shows three circles, each circumscribed by a regular polygon (shown from left to right, an equilateral triangle, a square, and a regular pentagon), all line segments are bisected by the points of tangency highlighted along the circles' circumferences:

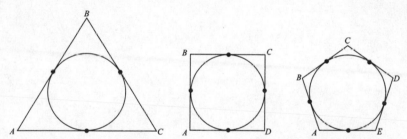

These two additional rules involving tangents allow for a variety of additional types of SAT questions.

Example:

In the figure below, $\overrightarrow{AB}$ passes through the center of circle O and $\overrightarrow{AC}$ is tangent to the circle at P.

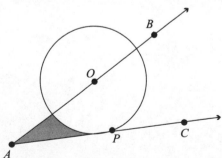

If the radius of the circle is 3 and $m\angle OAC = 30°$, what is the area of the shaded region?

(A) $\frac{3}{2}(3\sqrt{3} - \pi)$

(B) $\frac{1}{2}(3 - \sqrt{3}\pi)$

(C) $4\sqrt{3} - \pi$

(D) $\frac{2}{3}(3\pi - \sqrt{3})$

(E) $4\pi - \sqrt{3}$

Solution:

The correct answer is (A). Draw a radius from O to P. Since $\overleftrightarrow{AC}$ is tangent to the circle at P, $\overleftrightarrow{AC} \perp \overline{PO}$, and drawing the radius from O to P forms a right triangle ($\triangle AOP$), whose area = $\frac{1}{2}(3)(3\sqrt{3}) = \frac{9}{2}\sqrt{3}$. Since $m\angle OAP = 30°$, $m\angle OAP = 60°$ (one sixth the total number of degrees in the circle, 360), and hence the segment of the circle bound by $\angle OAC$ is one-sixth the circle's area, or $\frac{1}{6}\pi 3^2 = \frac{9}{6}\pi = \frac{3}{2}\pi$. To answer the question, subtract the area of this segment of the circle from the area of $\triangle AOP$: $\frac{9}{2}\sqrt{3} - \frac{3}{2}\pi = \frac{3}{2}(3\sqrt{3} - \pi)$.

Exercise 2

Work out each problem. Circle the letter that appears before your answer.

1. The figure below shows a regular pentagon tangent to circle O at five points.

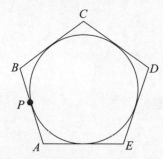

   If the perimeter of the pentagon is 10, what is the length of $\overline{AP}$?

2. In the figure below, $\overline{AC}$ is tangent to the circle at point B. The length of $\overline{BD}$ equals the diameter of the circle, whose center is O.

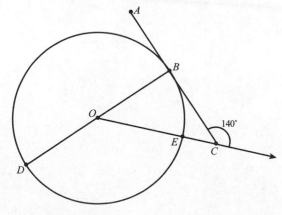

   What is the degree measure of minor arc DE ?

   (A) 40
   (B) 110
   (C) 120
   (D) 130
   (E) 220

3. In the figure below, C lies on the circumference of a circle with center O and radius 6.

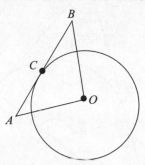

   If $m\angle BOA = 90°$ and $\overline{OA} \cong \overline{OB}$, what is the perimeter of $\triangle ABO$?

   (A) $6 + 12\sqrt{3}$
   (B) $12 + 12\sqrt{2}$
   (C) $18\sqrt{3}$
   (D) $24\sqrt{2}$
   (E) 36

4. The figure below shows an equilateral triangle ($\triangle ABC$) tangent to circle O at three points.

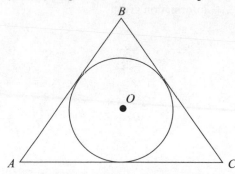

   If the perimeter of $\triangle ABC$ is 18, the area of circle $O =$

   (A) 2π
   (B) $\dfrac{5\pi}{2}$
   (C) $2\sqrt{2}\pi$
   (D) 3π
   (E) $2\pi\sqrt{3}$

5.   In the figure below, a circle with center O is tangent to $\overline{AB}$ at point D and tangent to $\overline{AC}$ at point C.

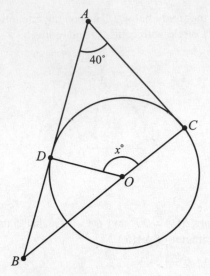

If $m\angle A = 40°$, then $x =$

(A)   140
(B)   145
(C)   150
(D)   155
(E)   It cannot be determined from the information given.

3. EQUATIONS AND GRAPHS OF LINES IN THE *XY*-PLANE

You can define any line in the standard *xy*-coordinate plane by the equation $y = mx + b$. In this equation, *m* is the slope of the line, *b* is the line's *y*-intercept (where the line crosses the *y* axis), and *x* and *y* are the coordinates of any point on the line. (Any (*x,y*) pair defining a point on the line can substitute for the variables *x* and *y*.)

You can determine the slope of a line from any two pairs of (*x,y*) coordinates. In general, if (x_1, y_1) and (x_2, y_2) lie on the same line, calculate the line's slope as follows (notice that you can subtract either pair from the other):

$$\text{slope } (m) = \frac{y_2 - y_1}{x_2 - x_1} \text{ or } \frac{y_1 - y_2}{x_1 - x_2}$$

Be careful to subtract corresponding values. For example, a careless test-taker calculating the slope might subtract y_1 from y_2 but subtract x_2 from x_1.

In the *xy*-plane:

- A line sloping *upward* from left to right has a positive slope (*m*). A line with a slope of 1 slopes upward from left to right at a 45° angle in relation to the *x*-axis. A line with a fractional slope between 0 and 1 slopes upward from left to right but at less than a 45° angle in relation to the *x*-axis. A line with a slope greater than 1 slopes upward from left to right at more than a 45° angle in relation to the *x*-axis.

- A line sloping *downward* from left to right has a negative slope (*m*). A line with a slope of –1 slopes downward from left to right at a 45° angle in relation to the *x*-axis. A line with a fractional slope between 0 and –1 slopes downward from left to right but at less than a 45° angle in relation to the *x*-axis. A line with a slope less than –1 (for example, –2) slopes downward from left to right at more than a 45° angle in relation to the *x*-axis.

- A *horizontal* line has a slope of zero ($m = 0$, and $mx = 0$)

- A *vertical* line has either an undefined or an indeterminate slope (the fraction's denominator is 0), so the *m*-term in the equation is ignored.

- *Parallel* lines have the same slope (the same *m*-term in the general equation).

- The slope of a line *perpendicular* to another is the negative reciprocal of the other line's slope. (The product of the two slopes is 1.) For example, a line with slope $\frac{3}{2}$ is perpendicular to a line with slope $-\frac{2}{3}$.

On the new SAT, a question involving the equation or graph of a line might ask you to apply one or more of the preceding rules in order to perform tasks such as:

- Identifying the slope of a line defined by a given equation (in which case you simply put the equation in the standard form $y = mx + b$, then identify the *m*-term.

- Determining the equation of a line, or just the line's slope (*m*) or *y*-intercept (*b*), given the coordinates of two points on the line.

- Determining the point at which two non-parallel lines intersect on the coordinate plane (in which case you determine the equation for each line, and then solve for *x* and *y* by either substitution or addition-subtraction)

- Recognizing the slope or the equation of a line based on the line's graph.

Example:

In the *xy*-plane, what is the slope of the line defined by the two points *P*(2,1) and *Q*(–3,4)?

(A)   –3

(B) $-\frac{5}{3}$

(C) $-\frac{3}{5}$

(D) $\frac{3}{5}$

(E)   3

Solution:

The correct answer is (C). Here are two ways to find the slope:

$$\text{slope }(m) = \frac{4-1}{-3-2} = \frac{3}{-5}$$

$$\text{slope }(m) = \frac{1-4}{2-(-3)} = \frac{-3}{5}$$

Example:

In the *xy*-plane, at what point along the *y*-axis does the line passing through points (5,–2) and (3,4) intersect that axis?

(A)   –8

(B) $-\frac{5}{2}$

(C)   3

(D)   7

(E)   13

Solution:

The correct answer is (E). The question asks for the line's *y*-intercept (the value of *b* in the general equation *y* = *mx* + *b*). First, determine the line's slope:

$$\text{slope }(m) = \frac{y_2 - y_1}{x_2 - x_1} = \frac{4-(-2)}{3-5} = \frac{6}{-2} = -3$$

In the general equation (*y* = *mx* + *b*), *m* = –3. To find the value of *b*, substitute either (*x*,*y*) value pair for *x* and *y*, then solve for *b*. Substituting the (*x*,*y*) pair (3,4):

$$y = -3x + b$$
$$4 = -3(3) + b$$
$$4 = -9 + b$$
$$13 = b$$

Example:

In the *xy*-plane, the (*x*,*y*) pairs (0,2) and (2,0) define a line, and the (*x*,*y*) pairs (–2,–1) and (2,1) define another line. At which of the following (*x*,*y*) points do the two lines intersect?

(A) $\left(\frac{4}{3}, \frac{2}{3}\right)$

(B) $\left(\frac{3}{2}, \frac{4}{3}\right)$

(C) $\left(-\frac{1}{2}, \frac{3}{2}\right)$

(D) $\left(\frac{3}{4}, -\frac{2}{3}\right)$

(E) $\left(-\frac{3}{4}, -\frac{2}{3}\right)$

Solution:

The correct answer is (A). For each line, formulate its equation by determining slope (m), then y-intercept (b). For the pairs (0,2) and (2,0):

$y = \left(\frac{0-2}{2-0}\right)x + b$ (slope $= -1$)

$0 = -2 + b$

$2 = b$

The equation for the line is $y = -x + 2$. For the pairs ($-2,-1$) and (2,1):

$y = \left(\frac{1-(-1)}{2-(-2)}\right)x + b$ (slope $= \frac{1}{2}$)

$1 = \frac{1}{2}(2) + b$

$0 = b$

The equation for the line is $y = \frac{1}{2}x$. To find the point of intersection, solve for x and y by substitution. For example:

$\frac{1}{2}x = -x + 2$

$\frac{3}{2}x = 2$

$x = \frac{4}{3}$

$y = \frac{2}{3}$

The point of intersection is defined by the coordinate pair ($\frac{4}{3}, \frac{2}{3}$).

Example:

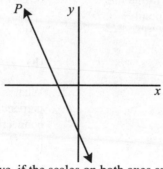

Referring to the *xy*-plane above, if the scales on both axes are the same, which of the following could be the equation of line P?

(A) $y = \frac{2}{5}x - \frac{5}{2}$

(B) $y = -\frac{5}{2}x + \frac{5}{2}$

(C) $y = \frac{5}{2}x - \frac{5}{2}$

(D) $y = \frac{2}{5}x + \frac{2}{5}$

(E) $y = -\frac{5}{2}x - \frac{5}{2}$

Solution:

The correct answer is (E). Notice that line P slopes downward from left to right at an angle greater than 45°. Thus, the line's slope (m in the equation $y = mx + b$) < -1. Also notice that line P crosses the y-axis at a negative y-value (that is, below the x-axis). That is, the line's y-intercept (b in the equation $y = mx + b$) is negative. Only choice (E) provides an equation that meets both conditions.

Exercise 3

Work out each problem. Circle the letter that appears before your answer.

1. In the xy-plane, what is the slope of the line described by the equation $2y = -9$?

    (A) $-\dfrac{9}{2}$

    (B) $-\dfrac{2}{9}$

    (C) 0

    (D) $\dfrac{9}{2}$

    (E) The slope is undefined.

2. In the xy-plane, what is the slope of a line that contains the points $(-1,4)$ and $(3,-6)$?

    (A) $-\dfrac{5}{2}$

    (C) -2

    (B) $-\dfrac{1}{2}$

    (D) 1

    (E) 2

3. In the xy-plane, what is the equation of the line with slope 3, if the line contains the point defined by the xy-coordinate pair $(-3,3)$?

    (A) $y = 3x - 3$
    (B) $y = 3x + 12$
    (C) $y = x + 6$
    (D) $y = -3x - 12$
    (E) $y = 6x - 6$

4. Referring to the xy-plane below, if the scales on both axes are the same, which of the following could be the equation of line P ?

    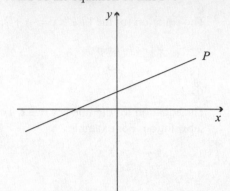

    (A) $y = \dfrac{2}{3}x - 6$

    (B) $y = \dfrac{3}{2}x - 6$

    (C) $y = -\dfrac{3}{2}x + 6$

    (D) $y = \dfrac{2}{3}x + 6$

    (E) $y = -\dfrac{2}{3}x - 6$

5. What is the equation of the line that is the perpendicular bisector of the line segment connecting points $(4,-2)$ and $(-3,5)$ in the xy-plane?

    (A) $y = -x + \dfrac{3}{2}$

    (B) $y = x + \dfrac{1}{2}$

    (C) $y = \dfrac{3}{2}x - 1$

    (D) $y = -x + 2$

    (E) $y = x + 1$

4. GRAPHS OF FUNCTIONS AND OTHER EQUATIONS—FEATURES AND TRANSFORMATIONS

On the new SAT, a question might show a graph of a quadratic function or other equation in the xy-plane, and then ask you to identify or recognize certain features of the graph—for example, minimum or maximum points on the graph. You might encounter the graph of a circle, an ellipse, a parabola, or even a trigonometric function (appearing as a wave). To answer these questions, you do not need to know the equations that define such graphs; simply apply your knowledge of the xy-coordinate system and, for some questions, function notation (see Chapter 15).

Example:

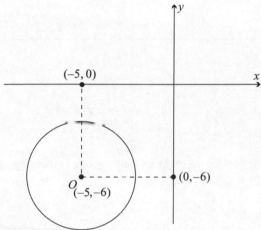

The figure above shows the graph of a certain equation in the xy-plane. The graph is a circle with center O and circumference 6π. At how many different values of y does $x = -7.5$?

(A)   0
(B)   1
(C)   2
(D)   4
(E)   Infinitely many

Solution:

The correct answer is (C). First, find the circle's radius from its circumference: $C = 6\pi = 2\pi r$; $r = 3$. Since the circle's center (O) lies at $(-5,-6)$, the minimum value in the domain of x is -8. In other words, the left-most point along the circle's circumference is at $(-8,-6)$, 3 units to the left of O. Thus, the graph of $x = -7.5$, which is a vertical line passing through $(-7.5, 0)$, intersects the circle at exactly two points. That is, when $x = -7.5$, there are two different corresponding values of y.

Other questions on the new SAT will involve *transformations* of linear and quadratic functions and the effect of transformations on the graphs of such functions. The function $f(x)$ is transformed by substituting an expression containing the variable x for x in the function — for example:

If $f(x) = 2x$, then $f(x + 1) = 2(x + 1)$, or $2x + 2$

Transforming a function alters the graph of the function in the xy-plane. The effect of a transformation might be any of the following:

* To move, or *translate*, the graph (either vertically, horizontally, or both) to another position in the plane

* To alter the slope of a line (in the case of a linear function)

* To alter the shape of a curve (in the case of a quadratic function)

For example, if $f(x) = x$, then $f(x + 1) = x + 1$. In the xy-plane, the graph of $f(x) = x$ (or $y = x$), is a line with slope 1 passing through the origin $(0,0)$. The effect of transforming $f(x)$ to $f(x + 1)$ on the graph of $f(x)$ is the translation of the line one unit upward. (The y-intercept becomes 1 instead of 0.) Remember: In determining the graph of a function in the xy-plane, use y to signify $f(x)$ and, conversely, use x to signify $f(y)$.

Example:

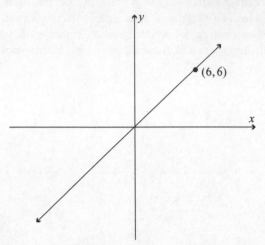

If $f(x) = x + 3$, then the line shown in the xy-plane above is the graph of

(A) $f(x)$

(B) $f(x - 6)$

(C) $f(x + 6)$

(D) $f(x + 3)$

(E) $f(x - 3)$

Solution:

The correct answer is (E). The figure shows the graph of the function $f(x) = x$ (or $y = x$). To determine which of the five answer choices transforms the original function $f(x) = x + 3$ to the function $f(x) = x$, substitute the variable expression in each choice, in turn, for x in the original function. Choice (E) is the only one that provides an expression that achieves this transformation:

$$f(x-3) = (x-3)+3$$
$$f(x-3) = x$$
$$y = x$$

To help you determine the effect of a function's transformation on the function's graph, you can tabulate some (x,y) pairs based on the new function, plot the points on the xy-plane, and then connect them.

Example:

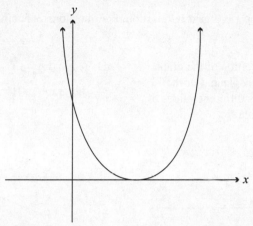

If $f(x) = x^2$, then the graph shown in the xy-plane above best represents which of the following functions?

(A) $f(-x)$
(B) $f(x - 1)$
(C) $f(x + 1)$
(D) $f(x^2 + 1)$
(E) $f(x^2 - 1)$

Solution:

The correct answer is (B). The figure shows the graph of $y = x^2$, but translated to the right. Substitute the variable expression given in each answer choice, in turn, for x in the function $f(x) = x^2$. Performing this task for choice (B) yields the equation $f(x) = (x - 1)^2$, or $y = (x - 1)^2$. Identify and plot some (x,y) pairs. (Since the vertex in the graph lies along the x-axis, let $x = 0$ in order to establish the vertex's coordinates.) Here are some (x,y) pairs for the equation $y = (x - 1)^2$:

$(0,1)(1,0)$, $(2,1)$, $(3,4)$, $(-1,4)$

Plotting these points in the xy-plane reveals a graph whose key features match those of the figure provided in the question.

Exercise 4

Work out each problem. Question 1 is a grid-in question. For questions 2–5, circle the letter that appears before your answer.

1. The figure below shows a portion of the graph of a certain function in the *xy*-plane. For the portion shown, at how many different values of *x* is | *f*(*x*)| at its maximum value?

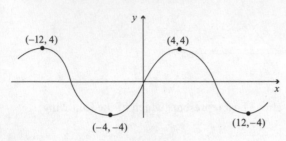

2. If *f*(*x*) = 2, then the line shown in the *xy*-plane below is the graph of

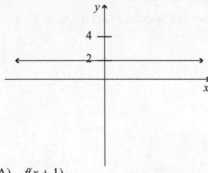

    (A) *f*(*x* + 1)
    (B) *f*(*x* – 1)
    (C) *f*(*x* + 2)
    (D) *f*(*x* – 2)
    (E) All of the above

3. If *f*(*x*) = 2*x* – 2, then which of the following is the graph of $f\left(\dfrac{x-2}{2}\right)$?

(A)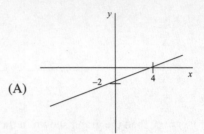

(B)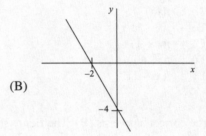

(C)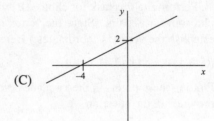

(D)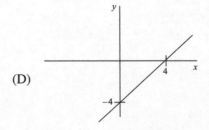

(E)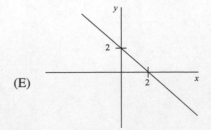

4. If $f(x) = (x - 1)^2 + 1$, what is the y-intercept of the graph of $f(x + 1)$ in the xy-plane?

    (A) −2
    (B) −1
    (C) 0
    (D) 1
    (E) 2

5. If $f(y) = -(y^2 + 1)$, then the graph shown in the xy-plane below best represents which of the following functions?

    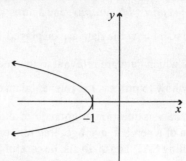

    (A) $f(-y)$
    (B) $f(y + 1)$
    (C) $f(y - 1)$
    (D) $f(y - 2)$
    (E) $f(y^2 - 2)$

5. DATA ANALYSIS

The new SAT includes questions involving the analysis of data displayed in graphical formats such as *tables*, *pie graphs*, *line charts*, *bar graphs*, and *scatter plots*. To answer a data-analysis question, you'll need to:

* Understand how the data are displayed

* Know which data are relevant to the question

* Know how to process the relevant data to solve the problem (answer the question correctly)

A data analysis question might require a simple arithmetic calculation (addition or subtraction) and/or a simple calculation of a percent, average, fraction, or ratio.

In handling SAT data analysis, be careful to read the question very carefully, so that you select the appropriate graphical data and perform the appropriate calculation — one that yields the answer to the precise question being asked. In analyzing a line chart, bar graph, or scatter plot (see the examples below), estimating number values in the display will suffice to answer the question correctly. To answer any data analysis question asking for an approximation, rounding off your calculations will suffice.

Example (Table):

WORLDWIDE SALES OF XYZ MOTOR COMPANY MODELS, 2002-03 MODEL YEAR

	Automobile Model		
	Basic	Standard	Deluxe
U.S. institutions	3.6	8.5	1.9
U.S. consumers	7.5	11.4	2.0
Foreign institutions	1.7	4.9	2.2
Foreign consumers	1.0	5.1	0.8

(Purchaser category)

Note: All numbers are in thousands.

According to the table above, of the total number of automobiles sold to U.S. and foreign institutions during the 2002–03 model year, which of the following most closely approximates the percent that were standard models?

(A)   24%

(B)   36%

(C)   41%

(D)   59%

(E)   68%

Solution:

The correct answer is (D). The total number of units sold to institutions = (3.6 + 8.5 + 1.9) + (1.7 + 4.9 + 2.2) = 22.8. The number of these units that were standard models = (8.5 + 4.9) = 13.4. To answer the question, divide 13.4 by 22.8 (round off the quotient): 22.8 ÷ 13.4 ≈ .59, or 59%.

Example (Pie Graph):

AREA OF WAREHOUSE UNITS A, B, C AND D
(AS PORTIONS OF TOTAL WAREHOUSE AREA)

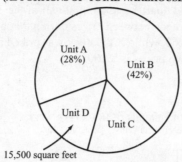

Unit A
(28%)

Unit B
(42%)

Unit D

Unit C

15,500 square feet

<u>Total:</u> 140,000 square feet
<u>Note:</u> Figure not drawn to scale.

Based on the data shown above, the combined area of Unit B and Unit D is approximately

(A)   51,000 square feet
(B)   57,500 square feet
(C)   70,000 square feet
(D)   74,500 square feet
(E)   108,000 square feet

Solution:

The correct answer is (D). The size of Unit B is 42% of 140,000 square feet, or about 59,000 square feet. Thus, the combined size of Unit B and Unit D is approximately 74,500 square feet.

Example (Line Chart):

PRICE OF COMMON STOCK OF
XYZ CORP. AND ABC CORP.
(YEAR X)

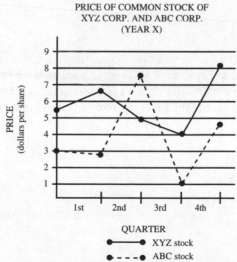

PRICE
(dollars per share)

QUARTER

●————● XYZ stock
●- - - -● ABC stock

Referring to the graph above, approximately what was the greatest dollar amount by which the share price of ABC common stock exceeded the share price of XYZ common stock at the same time during year X?

(A) $1.80
(B) $2.60
(C) $3.00
(D) $3.60
(E)   It cannot be determined from the information given.

Solution:

The correct answer is (B). You're looking for the point at which the dotted line (ABC's stock price) is furthest above the solid line (XYZ's stock price). The dotted line lies above the solid line only during the second half of the 2nd quarter and the first half of the 3rd quarter; the end of the 2nd quarter marks the greatest difference between prices during that period. At that time, ABC stock was priced at approximately $7.60, while XYZ stock was priced at approximately $5.00 per share. The difference between those two prices is $2.60.

Example (Bar Graph):

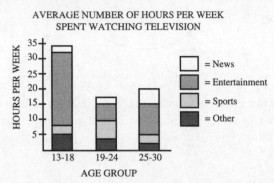

Referring to the data shown above, what is the approximate ratio of the average number of hours per week that the youngest age group spent watching entertainment to the average number of hours that the other two groups combined spent watching entertainment?

(A)   3:4
(B)   1:1
(C)   6:5
(D)   5:3
(E)   5:2

Solution:

The correct answer is (D). You're task here is to compare the size of the entertainment portion of the left-hand bar to the combined sizes of the same portion of the other to bars. Size up the ratio visually. The portion on the first chart is a bit larger than the other two combined, and so you're looking for a ratio that's greater than 1:1. Approximate the height of each three portions:

13–18 age group: 25 hours

19–24 age group: 5 hours

25–30 age group: 10 hours

The ratio in question is 25:15, or 5:3.

Example (Scatter Plot):

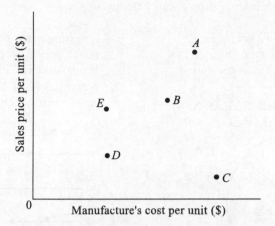

Companies A, B, C, D, and E all manufacturer and sell a similar product. The graph above compares manufacturing costs and sales prices per unit among the five companies. If all five companies have sold the same number of units, which company has earned the greatest profit from those sales?

(A)   A
(B)   B
(C)   C
(D)   D
(E)   E

Solution:

The correct answer is (E). Since the number of units sold was the same for all five companies, the greatest profit was earned by the company with the highest price-to-cost ratio. You can compare ratios by drawing a line segment from point *0* to each of the five plotted points. The segment with the steepest slope (vertical change divided by horizontal change) indicates the greatest price-to-cost ratio. Segment $\overline{OE}$ has the steepest slope, and hence company E earned the greatest profit.

Exercise 5

Work out each problem. Circle the letter that appears before your answer.

1. According to the data shown below, by approximately what amount did Division D's income exceed Division C's income during year X?

    WEBCO'S INCOME DURING YEAR X —
    DIVISIONS A, B, C, AND D

    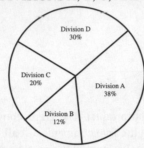

    INCOME
    (Total Income = $1,560,000)

    (A) $125,000
    (B) $127,000
    (C) $140,000
    (D) $156,000
    (E) $312,000

2. Among the years covered in the graph below, during the year in which aggregate awards of non-minority and minority funds was greatest, the dollar difference between non-minority and minority awards was approximately:

    STATE SCHOLARSHIP FUNDS AWARDED (1985-2000)
    ☐ Non-minority scholarship funds
    ▨ Minority scholarship funds

    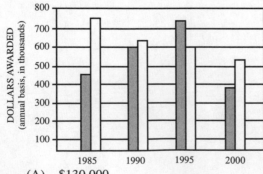

    (A) $130,000
    (B) $160,000
    (C) $220,000
    (D) $270,000
    (E) $400,000

3. Referring to the graph below, during the two-month period over which the average daily temperature in City X increased by the greatest percentage, City Y's highest daily temperature was approximately:

    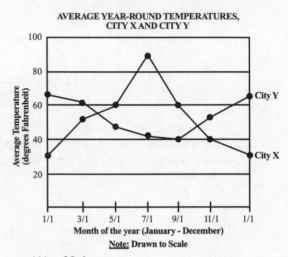

    (A) 38 degrees
    (B) 42 degrees
    (C) 52 degrees
    (D) 62 degrees
    (E) 66 degrees

Questions 4 and 5 are based on the following figure, which compares the race times of ten different cyclists, all of whom competed in the same two races (race 1 and race 2).

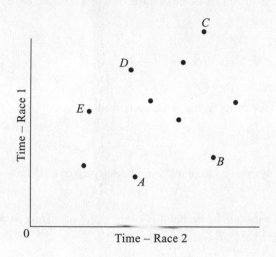

4. Among the five cyclists identified in the figure as A, B, C, D, and E, which had the fastest combined (total) race time for races 1 and 2?

(A) A
(B) B
(C) C
(D) D
(E) E

5. Considering the ten cyclists as a group, which of the following most closely approximates the ratio of the average time for race 1 to the average time for race 2?

(A) 1:2
(B) 2:3
(C) 1:1
(D) 3:2
(E) 2:1

6. PROBABILITY

The new SAT includes simple questions involving *probability*, which refers to the statistical chances, or "odds," of an event occurring (or not occurring). By definition, probability ranges from 0 to 1. (Probability is never negative, and it's never greater than 1.) You can express probability either as either a fraction or a percent. Here's the basic formula:

$$\text{Probability} = \frac{\text{number of ways the event can occur}}{\text{total number of possible occurrences}}$$

Example:

A standard deck of 52 playing cards contains 12 face cards. What is the probability of selecting a face card from such a deck?

Solution:

The correct answer is $\frac{12}{52}$, or $\frac{3}{13}$. There are 12 ways that a face card could be selected at random from the standard 52-card deck.

To calculate the probability of an event NOT occurring, just *subtract* the probability of the event occurring *from 1*. Referring to the preceding example, the probability of NOT selecting a face card would be $\frac{40}{52}$, or $\frac{10}{13}$. (Subtract $\frac{12}{52}$ from 1.)

An SAT probability problem might involve the probability of two *independent events* both occurring. Two events are "independent" if neither event affects the probability that the other will occur. Here are two general situations:

* The random selection of one object from *each of two groups* (for example, the outcome of throwing a pair of dice)

* The random selection of one object from a group, then *replacing* it and selecting again (as in a "second round" or "another turn" of a game)

To determine the probability of two independent events both occurring, *multiply* individual probabilities.

Example:

If you randomly select one letter from each of two sets: {A,B} and {C,D,E}, what is the probability of selecting A and C?

Solution:

The correct answer is $\frac{1}{6}$. The probability of selecting A from the set {A,B} is $\frac{1}{2}$, while the probability of selecting C from the set {C,D,E} is $\frac{1}{3}$. Hence, the probability of selecting A and C is $\frac{1}{2} \times \frac{1}{3}$, or $\frac{1}{6}$.

An SAT probability problem might be accompanied by a geometry figure or other figure that provides a visual display of the possibilities from which you are to calculate a probability.

Example:

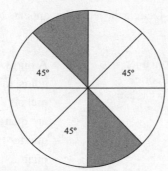

If a point is selected at random from the circular region shown above, what is the probability that the point will lie in a shaded portion of the circle?

Solution:

The correct answer is .25 (or $\frac{1}{4}$). The angles opposite each of the three 45° angles identified in the figure must also measure 45° each. Given a total of 360° in a circle, all of the eight small angles formed at the circle's center measure 45°, and hence all eight segments of the circle are congruent. The two shaded segments comprise $\frac{2}{8}$, or $\frac{1}{4}$ (.25) of the circle's area. The probability of selecting a point at random in a shaded area is also $\frac{1}{4}$ (or .25).

Exercise 6

Work out each problem. For questions 1–4, circle the letter that appears before your answer. Question 5 is a grid-in question.

1. If you randomly select one candy from a jar containing two cherry candies, two licorice candies, and one peppermint candy, what is the probability of selecting a cherry candy?

   (A) $\frac{1}{6}$

   (B) $\frac{1}{5}$

   (C) $\frac{1}{3}$

   (D) $\frac{2}{5}$

   (E) $\frac{3}{5}$

2. Patrons at a certain restaurant can select two of three appetizers—fruit, soup and salad—along with two of three vegetables—carrots, squash and peas. What is the probability that any patron will select fruit, salad, squash, and peas?

   (A) $\frac{1}{12}$

   (B) $\frac{1}{9}$

   (C) $\frac{1}{6}$

   (D) $\frac{1}{3}$

   (E) $\frac{1}{2}$

3. If one student is chosen randomly out of a group of seven students, then one student is again chosen randomly from the same group of seven, what is the probability that two different students will be chosen?

   (A) $\frac{36}{49}$

   (B) $\frac{6}{7}$

   (C) $\frac{19}{21}$

   (D) $\frac{13}{14}$

   (E) $\frac{48}{49}$

4. A piggy-bank contains a certain number of coins, of which 53 are dimes and 19 are nickels. The remainder of the coins in the bank are quarters. If the probability of selecting a quarter from this bank is $\frac{1}{4}$, how many quarters does the bank contain?

   (A) 16
   (B) 21
   (C) 24
   (D) 27
   (E) 30

5. The figure below shows two concentric circles, each divided into six congruent segments. The area of the large circle is exactly 3 times that of the smaller circle.

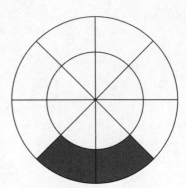

If a point is selected at random from the large circular region, what is the probability that the point will lie in a shaded portion of that circle?

RETEST

Answer questions 1–12. Question 9 is a "grid-in" (student-produced response) question; all other questions are standard multiple-choice. Try to answer questions 1 and 2 using trigonometry.

1. In the triangle shown below, what is the value of x ?

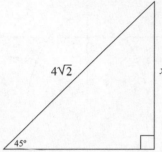

   (A) 2
   (B) $2\sqrt{2}$
   (C) 3
   (D) $\dfrac{5\sqrt{2}}{2}$
   (E) 4

2. Two planes depart at the same time from the same terminal, one traveling due north and the other due west, each on a straight flight path. When the shortest distance between the planes is 40 miles, one plane would need to turn 120° to either the left or right to point directly at the other plane. Which of the following most closely approximates the number of miles the faster of the two planes has traveled at this point in time?

   (A) 25
   (B) 30
   (C) 35
   (D) 40
   (E) 45

3. In the figure below, O_1 and O_2 are concentric circles and $\overline{AB}$ is tangent to O_1 at C.

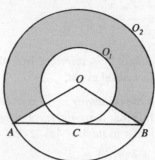

   If the radius of O_1 is r and the radius of O_2 is twice as long, what is the area of the shaded region?

   (A) $\frac{1}{2}\pi r^2$
   (B) πr^2
   (C) $\frac{3}{2}\pi r^2$
   (D) $2\pi r^2$
   (E) $3\pi r^2$

4. In the xy-plane below, if the scales on both axes are the same, which of the following could be the equation of l_1 ?

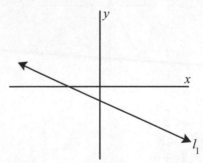

   (A) $y = \dfrac{2}{3}x - 3$
   (B) $y = -2x + 1$
   (C) $y = x + 3$
   (D) $y = -3x - \dfrac{2}{3}$
   (E) $y = -\dfrac{2}{3}x - 3$

5. In the *xy*-plane, if lines *a* and *b* intersect at point (5,–2) and lines *b* and *c* intersect at point (–3,3), what is the slope of line *b* ?

    (A) $-\dfrac{5}{2}$

    (B) $-\dfrac{5}{8}$

    (C) $-\dfrac{2}{5}$

    (D) $\dfrac{1}{2}$

    (E) It cannot be determined from the information given.

6. Which of the following is the equation of a straight line that has *y*-intercept 3 and is perpendicular to the line $4x - 2y = 6$?

    (A) $2y + 3x = -3$
    (B) $y + 3x = 2$
    (C) $2y - x = 6$
    (D) $y - 2x = 4$
    (E) $2y + x = 6$

7. If $f(x) = -\dfrac{1}{2}x$, then the line shown in the *xy*-plane below is the graph of

    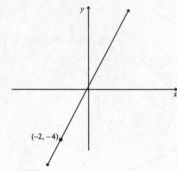

    (A) $f(x)$
    (B) $f(x - 3)$
    (C) $f(x + 3)$
    (D) $f(-4x)$
    (E) $f(x + 6)$

8. If $f(x) = 2x^2 + 2$, then the graph shown in the *xy*-plane below best represents which of the following functions?

    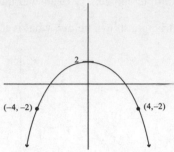

    (A) $f\left(\dfrac{1}{2}\right)$

    (B) $f\left(-\dfrac{1}{x}\right)$

    (C) $f(-2x)$

    (D) $f\left(-\dfrac{x}{4}\right)$

    (E) $f\left(-\dfrac{x}{2}\right)$

9. Based on the data shown below, how many chickens at Hill Farm laid 10 eggs from June 1st through June 7th?

    EGGS PRODUCED AT HILL FARM
    (WEEK OF JUNE 1 – JUNE 7)

NUMBER OF EGGS PER CHICKEN	10	9	8	7	6	5	4	3	0–2
NUMBER OF CHICKENS	?	2	4	5	3	2	0	2	0
TOTAL NUMBER OF EGGS AT HILL FARM: 179									

10. According to the data shown below, during what year was the dollar amount of Country Y's exports approximately twice that of Country X's imports?

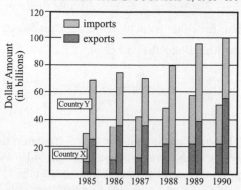

IMPORTS AND EXPORTS FOR
COUNTRY X AND COUNTRY Y, 1985–1990

(A) 1985
(B) 1987
(C) 1988
(D) 1989
(E) 1990

11. A bag of marbles contains twice as many red marbles as blue marbles, and twice as many blue marbles as green marbles. If these are the only colors of marbles in the bag, what is the probability of randomly picking a blue marble from the bag?

(A) $\frac{1}{6}$

(B) $\frac{2}{9}$

(C) $\frac{1}{4}$

(D) $\frac{2}{7}$

(E) $\frac{1}{3}$

12. The figure below shows two T-shaped cardboard pieces, both to be folded into a pair of cube-shaped dice.

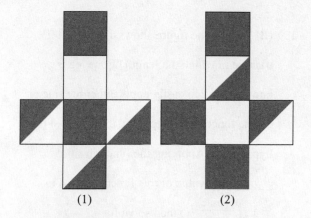

(1)   (2)

On a fair throw of both dice, what is the probability that NEITHER die will show either a solid white or solid black surface facing up?

(A) $\frac{1}{6}$

(B) $\frac{1}{5}$

(C) $\frac{1}{4}$

(D) $\frac{1}{3}$

(E) $\frac{2}{5}$

SOLUTIONS TO PRACTICE EXERCISES

Diagnostic Test

1. **(B)** Since the figure shows a 45°-45°-90° triangle in which the length of one leg is known, you can easily apply either the sine or cosine function to determine the length of the hypotenuse. Applying the function $\sin 45° = \frac{\sqrt{2}}{2}$, set the value of this function equal to $\frac{5}{x}$ $\left(\frac{\text{opposite}}{\text{hypotenuse}}\right)$, then solve for x:

   $\frac{\sqrt{2}}{2} = \frac{5}{x}$; $\sqrt{2}x = 10$; $x = \frac{10}{\sqrt{2}} = \frac{10\sqrt{2}}{2} = 5\sqrt{2}$.

2. **(D)** Since the figure shows a 30°-60°-90° triangle, you can easily apply either the sine or the cosine function to determine the length of the hypotenuse. Applying the function $\sin 30° = \frac{1}{2}$, set the value of this function equal to $\frac{4}{x}$ $\left(\frac{\text{opposite}}{\text{hypotenuse}}\right)$, then solve for x: $\frac{1}{2} = \frac{4}{x}$; $x = 8$.

3. **(C)** Since the hexagon is regular (all sides are congruent), the area of ΔAOP in the following figure is $\sqrt{3}$ — one sixth the area of the hexagon.

   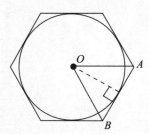

   ΔAOP is equilateral; hence you can divide it into two $1 : \sqrt{3} : 2$ triangles, as shown in the figure. Since the common leg, whose length is $\sqrt{3}$, is also the circle's radius, the circle's circumference must be $2\pi\sqrt{3}$.

4. **(E)** The line's slope $(m) = \frac{y_2 - y_1}{x_2 - x_1} = \frac{1-3}{4-2} = \frac{-2}{2} = -1$. Substitute the (x,y) pair for either point to define the equation of the line. Using the pair (2,3):

   $y = -x + b$
   $3 = -2 + b$
   $5 = b$

   The line's equation is $y = -x + 5$. To determine which of the five answer choices provides a point that also lies on this line, plug in the values of x and y provided each answer choice, in turn. Only choice (E) provides a solution to the equation: $-1 = -6 + 5$.

5. **(B)** The slope of $\overline{AB} = \frac{y_2 - y_1}{x_2 - x_1} = \frac{3-(-3)}{4-(-4)} = \frac{6}{8} = \frac{3}{4}$. The slope of the line perpendicular to $\overline{AB}$ is the negative reciprocal of $\frac{3}{4}$, which is $-\frac{4}{3}$.

6. **(D)** Given any two xy-coordinate points, a line's slope $m = \frac{y_1 - y_2}{x_1 - x_2}$. Accordingly, $\frac{1}{3} = \frac{5-(-3)}{a-2}$. Simplify, then cross-multiply to solve for a:

   $\frac{1}{3} = \frac{8}{a-2}$
   $a - 2 = (3)(8)$
   $a - 2 = 24$
   $a = 26$

7. **(B)** By visual inspection, you can see that the maximum value of y is 2, and this value occurs only once in the set of y-values — when $x = 0$.

8. **(B)** Substitute the variable expression provided in each answer choice, in turn, for x in the function, and you'll find that only choice (B) provides an expression that transforms the function into one whose graph matches the one in the figure: $f(x + 1) = x + 1$. To confirm that the line in the figure is in fact the graph of $y = x + 1$, substitute the two (x,y) pairs plotted along the line for x and y in the equation. The equation holds for both pairs: $(3) = (2) + 1$; $(-2) = (-3) + 1$.

9. The correct answer is 25. The total number of bowlers in the league is 240, which is the total of the six numbers in the frequency column. The number of bowlers whose averages fall within the interval 161–200 is 60 $(37 + 23)$. These 60 bowlers account for $\frac{60}{240}$, or 25%, of the league's bowlers.

10. **(C)** In the scatter plot, B and D are further to the left *and* further up than all of the other three points (A, C, and E), which means that City B and City D receive less rainfall but higher temperatures than any of the other three cities. Statement (C) provides an accurate general statement, based on this information.

11. **(D)** Of six marbles altogether, two are blue. Hence, the chances of drawing a blue marble are 2 in 6, or 1 in 3, which can be expressed as the fraction $\frac{1}{3}$.

12. The correct answer is 3/16. Given that the ratio of the large circle's area to the small circle's area is 2:1, the small circle must comprise 50% of the total area of the large circle. The shaded areas comprise $\frac{3}{8}$ the area of the small circle, and so the probability of randomly selecting a point in one of these three regions is $\frac{3}{8} \times \frac{1}{2} = \frac{3}{16}$.

Exercise 1

1. **(B)** Since the figure shows a 45°-45°-90° triangle in which the length of one leg is known, you can easily apply either the sine or cosine function to determine the length of the hypotenuse. Applying the function $\sin 45° = \frac{\sqrt{2}}{2}$, set the value of this function equal to $\frac{x}{7} \left(\frac{\text{opposite}}{\text{hypotenuse}} \right)$, then solve for x:
$\frac{\sqrt{2}}{2} = \frac{x}{7}$; $2x = 7\sqrt{2}$; $x = \frac{7\sqrt{2}}{2}$.

2. **(C)** Since the figure shows a 30°-60°-90° triangle, you can easily apply either the sine or the cosine function to determine the length of the hypotenuse. Applying the function $\cos 60° = \frac{1}{2}$, set the value of this function equal to $\frac{x}{10} \left(\frac{\text{adjacent}}{\text{hypotenuse}} \right)$, then solve for x: $\frac{1}{2} = \frac{x}{10}$; $2x = 10$; $x = 5$.

3. **(B)** The question describes the following 30°-60°-90° triangle:

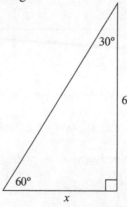

Since the length of one leg is known, you can easily apply the tangent function to determine the length of the other leg (x). Applying the function $\tan 30° = \frac{\sqrt{3}}{3}$, set the value of this function equal to $\frac{x}{6} \left(\frac{\text{opposite}}{\text{adjacent}} \right)$, then solve for x: $\frac{\sqrt{3}}{3} = \frac{x}{6}$; $3x = 6\sqrt{3}$; $x = 2\sqrt{3}$.

4. **(D)** The area of a triangle = $\frac{1}{2} \times$ base $\times$ height. Since the figure shows a 30°-60°-90° triangle with base 3, you can easily apply the tangent function to determine the height (the vertical leg). Applying the function $\tan 60° = \sqrt{3}$, let x equal the triangle's height, set the value of this function equal to $\frac{x}{3}$ $\left(\frac{\text{opposite}}{\text{adjacent}}\right)$, then solve for x: $\frac{\sqrt{3}}{1} = \frac{x}{3}$; $x = 3\sqrt{3}$. Now you can determine the triangle's area:

$\frac{1}{2} \times 3 \times 3\sqrt{3} = \frac{9}{2}\sqrt{3}$, or $\frac{9\sqrt{3}}{2}$.

5. **(A)** The two tracks form the legs of a right triangle, the hypotenuse of which is the shortest distance between the trains. Since the trains traveled at the same speed, the triangle's two legs are congruent (equal in length), giving us a 1:1:$\sqrt{2}$ with angles 45°, 45°, and 90°, as the next figure shows:

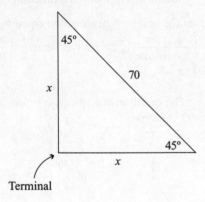

Terminal

To answer the question, you can solve for the length of either leg (x) by applying either the sine or cosine function. Applying the function $\cos 45° = \frac{\sqrt{2}}{2}$, set the value of this function equal to $\frac{\sqrt{2}}{x}$ $\left(\frac{\text{adjacent}}{\text{hypotenuse}}\right)$, then solve for x:

$\frac{\sqrt{2}}{2} = \frac{x}{70}$; $2x = 70\sqrt{2}$; $x = 35\sqrt{2}$. The question asks for an *approximate* distance in miles.

Using 1.4 as the approximate value of $\sqrt{2}$: $x \approx$

$(35)(1.4) = 49$.

Exercise 2

1. The correct answer is 1. $\overline{AB}$ is tangent to $\overline{PO}$; therefore, $\overline{AB} \perp \overline{PO}$. Since the pentagon is regular (all sides are congruent), P bisects $\overline{AB}$. Given that the perimeter of the pentagon is 10, the length of each side is 2, and hence $AP = 1$.

2. **(D)** Since $\overline{AC}$ is tangent to the circle, $m\angle ORC = 90°$. $\angle BCO$ is supplementary to the 140° angle shown; thus, $m\angle BCO = 40°$ and, accordingly, $m\angle BOE = 50°$. Since $\angle BOE$ and $\angle DOE$ are supplementary, $m\angle DOE = 130°$. (This angle measure defines the measure of minor arc DE.)

3. **(B)** Since $\overline{AB}$ is tangent to circle O at C, you can draw a radius of length 6 from O to C, forming two congruent 45°-45°-90° triangles ($\triangle ACO$ and $\triangle BCO$), each with sides in the ratio 1:1:$\sqrt{2}$. Thus, $OA = OB = 6\sqrt{2}$, and $AC = CB = 6$. The perimeter of $\triangle ABO = 12 + 6\sqrt{2} + 6\sqrt{2} = 12 + 12\sqrt{2}$.

4. **(D)** To find the circle's area, you must first find its radius. Draw a radius from O to any of the three points of tangency, and then construct a right triangle—for example, $\triangle ABC$ in the following figure:

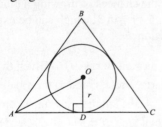

Since $m\angle BAC = 60°$, $m\angle OAD = 30°$, and $\triangle AOD$ is a 1:$\sqrt{3}$:2 triangle. Given that the perimeter of $\triangle ABC$ is 18, $AD = 3$. Letting $x = OD$: $\frac{3}{x} = \frac{\sqrt{3}}{1}$; $\sqrt{3}x = 3$; $x = \frac{3}{\sqrt{3}}$, or $\sqrt{3}$. The circle's radius = $\sqrt{3}$. Hence, its area = $\pi\left(\sqrt{3}\right)^2 = 3\pi$.

5. **(A)** Since $\overline{AC}$ is tangent to the circle, $\overline{AC} \perp \overline{BC}$. Accordingly, $\triangle ABC$ is a right triangle, and $m\angle B = 50°$. Similarly, $\overline{AB} \perp \overline{DO}$, $\triangle DBO$ is a right triangle, and $m\angle DOB = 40°$. $\angle DOC$ (the angle in question) is supplementary to $\angle DOB$. Thus, $m\angle DOB = 140°$. ($x = 140$.)

Exercise 3

1. **(C)** For all values of x, $y = -\frac{9}{2}$. Thus, the equation describes a horizontal line with y-intercept $-\frac{9}{2}$. The slope of a horizontal line is 0 (zero).

2. **(A)** Apply the formula for determining a line's slope (m):

$$m = \frac{y_2 - y_1}{x_2 - x_1} = \frac{-6 - 4}{3 - (-1)} = \frac{-10}{4} = -\frac{5}{2}$$

3. **(B)** In the general equation $y = mx + b$, the slope (m) is given as 3. To determine b, substitute -3 for x and 3 for y, then solve for b: $3 = 3(-3) + b$; $12 = b$.

4. **(D)** Line P slopes upward from left to right at an angle less than 45°. Thus, the line's slope (m in the equation $y = mx + b$) is a positive fraction less than 1. Also, line P crosses the y-axis at a positive y-value (above the x-axis). Thus, the line's y-intercept (b in the equation $y = mx + b$) is positive. Only choice (D) provides an equation that meets both conditions.

5. **(E)** First, find the midpoint of the line segment, which is where it intersects its perpendicular bisector. The midpoint's x-coordinate is $\frac{4-3}{2} = \frac{1}{2}$, and its y-coordinate is $\frac{-2+5}{2} = \frac{3}{2}$. Next, determine the slope of the line segment: $\frac{5-(-2)}{-3-4} = \frac{7}{-7} = -1$. Since the slope of the line segment is -1, the slope of its perpendicular bisector is 1. Plug (x,y) pair $(\frac{1}{2}, \frac{3}{2})$ and slope (m) 1 into the standard form of the equation for a line ($y = mx + b$), then solve for b (the y-intercept):

$$\frac{3}{2} = (1)\left(\frac{1}{2}\right) + b$$

$$1 = b$$

You now know the equation of the line: $y = x + 1$.

Exercise 4

1. The correct answer is 4. For every value of x, $f(x)$ is the corresponding y-value. By visual inspection, you can see that the maximum y-value is 4 and that the graph attains this value twice, at $(-8,4)$ and $(4,4)$. Similarly, the minimum value of y is -4 and the graph attains this value twice, at $(-4,-4)$ and $(8,-4)$; in both instances, the absolute value of y is 4. Thus, the absolute value of y is at its maximum at *four* different x-values.

2. **(E)** The figure shows the graph of $y = 2$. For any real number x, $f(x) = 2$. Thus, regardless of what number is added to or subtracted from x, the result is still a number whose function is 2 ($y = 2$).

3. **(D)** To determine the features of the transformed line, substitute $\frac{x-2}{2}$ for x in the function:

$$f\left(\frac{x-2}{2}\right) = \left(\frac{x-2}{2}\right) - 2 = x - 2 - 2 = x - 4$$

The correct figure should show the graph of the equation $y = x - 4$. Choice (D) shows the graph of a line with slope 1 and y-intercept -4, which matches the features of this equation. No other answer choice provides a graph with both these features.

4. **(D)** Substitute $(x + 1)$ for x in the function:

$$f(x + 1) = [(x + 1) - 1]^2 + 1 = x^2 + 1$$

In the xy-plane, the equation of $f(x + 1)$ is $y = x^2 + 1$. To find the y-intercept of this equation's graph, let $x = 0$, then solve for y:

$$y = (0)^2 + 1 = 1$$

5. **(A)** The graph of $x = -(y^2)$ is a parabola opening to the left with vertex at the origin $(0,0)$. The function $f(y) = -(y^2 + 1)$ is equivalent to $f(y) = -y^2 - 1$, the graph of which is the graph of $x = -(y^2)$, except translated one unit to the left, as the figure shows. [Since $(-y)^2 = y^2$ for any real number y, substituting $-y$ for y in the function $f(y) = -(y^2 + 1)$ does *not* transform the function in any way.]

Exercise 5

1. **(D)** Division D's income accounted for 30% of $1,560,000, or $468,000. Income from Division C was 20% of $1,560,000, or $312,000. To answer the question, subtract: $468,000 – $312,000 = $156,000.

2. **(A)** Visual inspection reveals that the aggregate amount awarded in 1995 exceeded that of any of the other 3 years shown. During that year, minority awards totaled approximately $730,000 and non-minority awards totaled approximately $600,000. The difference between the two amounts is $130,000.

3. **(E)** The two greatest two-month percent increases for City X were from 1/1 to 3/1 and from 5/1 to 7/1. Although the temperature increased by a greater *amount* during the latter period, the *percent* increase was greater from 1/1 to 3/1:

   January–March: from 30 degrees to 50 degrees, a 66% increase

   May–July: from 60 degrees to 90 degrees, a 50% increase

   During the period from 1/1 to 3/1, the highest daily temperature for City Y shown on the chart is appoximately 66 degrees.

4. **(A)** To answer the question, you can add together the "rise" (vertical distance) and the "run" (horizontal distance) from point *O* to each of the five lettered points (A–E). The shortest combined length represents the fastest combined (total) race time. Or, you can draw a line segment from point *O* to each of the five points—the shortest segment indicating the fastest combined time. As you can see, $\overline{OA}$ is the shortest segment, showing that cyclist A finished the two races in the fastest combined time.

5. **(C)** You can approximate the (race 1):(race 2) time ratio for the ten cyclists as a group by drawing a ray extending from point *O* through the "middle" of the cluster of points—as nearly as possible. Each of the five answer choices suggests a distinct slope for the ray. Choice (C) suggests a ray with slope 1 (a 45° angle), which does in fact appear to extend through the middle of the points:

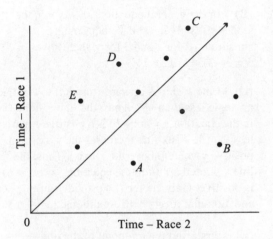

(Although six points are located above the ray, while only four are located below the ray, the ones below the ray, as a group, are further from the ray; so the overall distribution of values is fairly balanced above versus below the ray.) Any ray with a significantly flatter slope (answer choice A or B) or steeper slope (answer choice D or E) would not extend through the "middle" of the ten points and therefore would not indicate an accurate average (race 1):(race 2) ratio.

Exercise 6

1. **(D)** There are two ways among five possible occurrences that a cherry candy will be selected. Thus, the probability of selecting a cherry candy is $\frac{2}{5}$.

2. **(B)** In each set are three distinct member pairs. Thus the probability of selecting any pair is one in three, or $\frac{1}{3}$. Accordingly, the probability of selecting fruit and salad from the appetizer menu along with squash and peas from the vegetable menu is $\frac{1}{3} \times \frac{1}{3} = \frac{1}{9}$.

3. **(E)** You must first calculate the chances of picking *the same student twice,* by multiplying together the two individual probabilities for the student: $\frac{1}{7} \times \frac{1}{7} = \frac{1}{49}$. The probability of picking the same student twice, added to the probability of not picking the same student twice, equals 1. So to answer the question, subtract $\frac{1}{49}$ from 1.

4. **(C)** Let $x =$ the number of quarters in the bank (this is the numerator of the probability formula's fraction), and let $x + 72 =$ the total number of coins (the fraction's denominator). Solve for x:
$$\frac{1}{4} = \frac{x}{x+72}$$
$$x + 72 = 4x$$
$$72 = 3x$$
$$24 = x$$

5. The correct answer is 1/6. Given that the ratio of the large circle's area to the small circle's area is 3:1, the area of the "ring" must be twice that of the small circle. Hence the probability of randomly selecting a point in the outer ring is $\frac{2}{3}$. The shaded area accounts for $\frac{1}{4}$ of the ring, and so the probability of selecting a point in the shaded area is $\frac{2}{3} \times \frac{1}{4} = \frac{2}{12} = \frac{1}{6}$.

Retest

1. **(E)** Since the figure shows a 45°-45°-90° triangle in which the length of the hypotenuse is known, you can easily apply either the sine or cosine function to determine the length of either leg. Applying the function $\cos 45° = \frac{\sqrt{2}}{2}$, set the value of this function equal to $\frac{\sqrt{2}}{x} \left(\frac{\text{adjacent}}{\text{hypotenuse}} \right)$, then solve for x:
$$\frac{\sqrt{2}}{2} = \frac{x}{4\sqrt{2}} \; ; 2x = \left(4\sqrt{2}\right)\left(\sqrt{2}\right) \; ; 2x = 8 \; ; x = 4 .$$

2. **(C)** The two flight paths form the legs of a right triangle, the hypotenuse of which is the shortest distance between the trains (40 miles). As the next figure shows, a 120° turn to either the left or right allows for two scenarios (point T is the terminal):

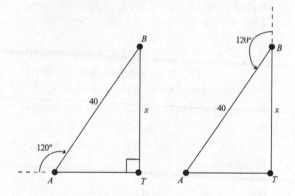

As the figures show, the two flight paths, along with a line segment connecting the two planes, form a 30°-60°-90° triangle with sides in the ratio $1 : \sqrt{3} : 2$. To answer the question, solve for the length of the longer leg (x), which is opposite the 60° angle. One way to solve for x is by applying either the sine or cosine function. Applying the function $\sin 60° = \frac{\sqrt{3}}{2}$, set the value of this function equal to $\frac{x}{40} \left(\frac{\text{opposite}}{\text{hypotenuse}} \right)$, then solve for x:

$\frac{\sqrt{3}}{2}=\frac{x}{40}$; $2x=40\sqrt{3}$; $x=20\sqrt{3}$. The question asks for an *approximate* distance in miles.

Using 1.7 as the approximate value of $\sqrt{3}$: $x\approx$ (20)(1.7) = 34.

3. **(D)** The entire area between the two circles is the area of the larger minus the area of the smaller. Letting that area equal A:

$$A=\pi(2r)^2-\pi r^2$$
$$=4\pi r^2-\pi r^2$$
$$=3\pi r^2$$

Drawing a line segment from C to O forms two right triangles, each with hypotenuse $2r$. Since $OC = r$, by the Pythagorean Theorem, the ratios among the triangle's sides are $1:\sqrt{3}:2$, with corresponding angle ratios $90°:60°:30°$. $\angle A$ and $\angle B$ each $= 30°$. Accordingly, interior $\angle AOB$ measures $120°$, or one third the degree measure of the circle. Hence, the area of the shaded region is two thirds of area A and must equal $2\pi r^2$.

4. **(E)** The line shows a negative y-intercept (the point where the line crosses the vertical axis) and a negative slope less than -1 (that is, slightly more horizontal than a 45° angle). In equation (E), $-\frac{2}{3}$ is the slope and -3 is the y-intercept. Thus, equation (E) matches the graph of the line.

5. **(B)** Points $(5,-2)$ and $(-3,3)$ are two points on line b. The slope of b is the change in the y-coordinates divided by the corresponding change in the x-coordinate:

$$m_b=\frac{3-(-2)}{-3-5}=\frac{5}{-8}, \text{ or } -\frac{5}{8}$$

6. **(E)** Put the equation given in the question into the form $y = mx + b$:

$$4x-2y=6$$
$$2y=4x-6$$
$$y=2x-3$$

The line's slope (m) is 2. Accordingly, the slope of a line perpendicular to this line is $-\frac{1}{2}$. Given a y-intercept of 3, the equation of the perpendicular line is $y=-\frac{1}{2}x+3$. Reworking this equation to match the form of the answer choices yields $2y + x = 6$.

7. **(D)** The figure shows the graph of $y = 2x$, whose slope (2) is twice the negative reciprocal of $-\frac{1}{2}$, which is the slope of the graph of $f(x) = -\frac{1}{2}x$. You obtain this slope by substituting $-4x$ for x in the function: $f(-4x) = -\frac{1}{2}(-4x) = 2x$.

8. **(E)** Substitute the variable expression given in each answer choice, in turn, for x in the function $f(x) = -2x^2 + 2$. Substituting $-\frac{x}{2}$ (given in choice E) for x yields the equation $y=-\frac{x^2}{4}+2$:

$$f\left(-\frac{x}{2}\right)=(-2)\left(-\frac{x}{2}\right)^2+2=(-2)\left(\frac{x^2}{4}\right)+2=$$
$$-\frac{2x^2}{8}+2, \text{ or } -\frac{x^2}{4}+2$$

The graph of $y=-\frac{x^2}{4}$ is a downward opening parabola with vertex at the origin (0,0). The figure shows the graph of that equation, except translated 2 units up. To confirm that (E) is the correct choice, substitute the (x,y) pairs $(-4,-2)$ and $(4,-2)$, which are shown in the graph, for x and y in the equation $y=-\frac{x^2}{4}+2$, and you'll find that the equation holds for both value pairs.

9. The correct answer is 6. By multiplying the number of chickens by the number of eggs they lay per week, then adding together the products, you can find the number of eggs laid by chickens laying 9 or fewer eggs per week:

(2)(9) + (4)(8) + (5)(7) + (3)(6) + (2)(5) + (0)(4) + (2)(3) = 119 eggs.

To find the number of chickens that laid 10 eggs during the week, subtract 119 from 179 (the total number of eggs): $179 - 119 = 60$. Then divide 60 by 10 to get 6 chickens.

10. **(E)** For each year, visually compare the difference in height between Country X's white bar and Country Y's dark bar. (For each year, the left-hand bars represent data for Country X, while the right-hand bars represent data for Country Y.) A quick inspection reveals that only for the year 1990 is Country Y's dark bar approximately twice the height of Country X's white bar. Although you don't need to determine dollar amounts, during 1990, Country Y's imports totaled about $55 million, while Country X's exports totaled about $28 million.

11. **(D)** Regardless of the number of marbles in the bag, the red : blue : green marble ratio is 4:2:1. As you can see, blue marbles account for $\frac{2}{7}$ of the total number of marbles. Thus, the probability of picking a blue marble is $\frac{2}{7}$.

12. **(A)** The probability that the left-hand die will NOT show a solid face is 3 in 6, or $\frac{1}{2}$. The probability that the right-hand die will NOT show a solid face is 2 in 6, or $\frac{1}{3}$. To calculate the combined probability of these two independent events occurring, multiply: $\frac{1}{2} \times \frac{1}{3} = \frac{1}{6}$.

Practice Tests

PRACTICE TEST A
Answer Sheet

Directions: For each question, darken the oval that corresponds to your answer choice. Mark only one oval for each question. If you change your mind, erase your answer completely.

Section 1

1. Ⓐ Ⓑ Ⓒ Ⓓ Ⓔ
2. Ⓐ Ⓑ Ⓒ Ⓓ Ⓔ
3. Ⓐ Ⓑ Ⓒ Ⓓ Ⓔ
4. Ⓐ Ⓑ Ⓒ Ⓓ Ⓔ
5. Ⓐ Ⓑ Ⓒ Ⓓ Ⓔ
6. Ⓐ Ⓑ Ⓒ Ⓓ Ⓔ
7. Ⓐ Ⓑ Ⓒ Ⓓ Ⓔ

8. Ⓐ Ⓑ Ⓒ Ⓓ Ⓔ
9. Ⓐ Ⓑ Ⓒ Ⓓ Ⓔ
10. Ⓐ Ⓑ Ⓒ Ⓓ Ⓔ
11. Ⓐ Ⓑ Ⓒ Ⓓ Ⓔ
12. Ⓐ Ⓑ Ⓒ Ⓓ Ⓔ
13. Ⓐ Ⓑ Ⓒ Ⓓ Ⓔ
14. Ⓐ Ⓑ Ⓒ Ⓓ Ⓔ

15. Ⓐ Ⓑ Ⓒ Ⓓ Ⓔ
16. Ⓐ Ⓑ Ⓒ Ⓓ Ⓔ
17. Ⓐ Ⓑ Ⓒ Ⓓ Ⓔ
18. Ⓐ Ⓑ Ⓒ Ⓓ Ⓔ
19. Ⓐ Ⓑ Ⓒ Ⓓ Ⓔ
20. Ⓐ Ⓑ Ⓒ Ⓓ Ⓔ
21. Ⓐ Ⓑ Ⓒ Ⓓ Ⓔ

22. Ⓐ Ⓑ Ⓒ Ⓓ Ⓔ
23. Ⓐ Ⓑ Ⓒ Ⓓ Ⓔ
24. Ⓐ Ⓑ Ⓒ Ⓓ Ⓔ
25. Ⓐ Ⓑ Ⓒ Ⓓ Ⓔ

Section 2

Note: Only the answers entered on the grid are scored. Handwritten answers at the top of the column are not scored.

313

PRACTICE TEST A

Section 1

25 Questions

Time: 30 Minutes

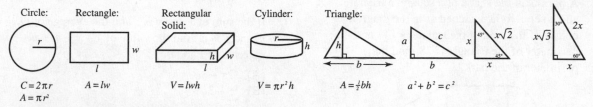

Circle: Rectangle: Rectangular Solid: Cylinder: Triangle:

$C = 2\pi r$
$A = \pi r^2$
$A = lw$
$V = lwh$
$V = \pi r^2 h$
$A = \frac{1}{2}bh$
$a^2 + b^2 = c^2$

The number of degrees of arc in a circle is 360.
The measure in degrees of a straight angle is 180.
The sum of the measures in degrees of the angles of a triangle is 180.

1. If 20% of a number is 8, what is 25% of the number?

   (A) 2
   (B) 10
   (C) 12
   (D) 11
   (E) 15

2. If $x + 3$ is a multiple of 3, which of the following is not a multiple of 3?

   (A) x
   (B) $x + 6$
   (C) $6x + 18$
   (D) $2x + 6$
   (E) $3x + 5$

3. In the figure below, $AB = AC$. Then $x =$

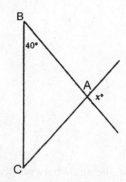

   (A) 40°
   (B) 80°
   (C) 100°
   (D) 60°
   (E) 90°

4. $\left(\frac{2}{5} \div \frac{2}{3}\right) + \left(\frac{1}{2} - \frac{1}{10}\right) =$

   (A) $-\frac{1}{10}$
   (B) $-\frac{1}{7}$
   (C) $\frac{19}{15}$
   (D) $\frac{1}{5}$
   (E) 1

5. The toll on the Islands Bridge is $1.00 for car and driver and $.75 for each additional passenger. How many people were riding in a car for which the toll was $3.25?

   (A) 2
   (B) 3
   (C) 4
   (D) 5
   (E) none of these

6. If $y^3 = 2y^2$ and $y \neq 0$, then y must be equal to

   (A) 1
   (B) $\frac{1}{2}$
   (C) 2
   (D) 3
   (E) −1

7. If x and y are negative integers and $x - y = 1$, what is the least possible value for xy?

   (A) 0
   (B) 1
   (C) 2
   (D) 3
   (E) 4

8. A park is in the shape of a square, a triangle, and a semicircle, attached as in the diagram below. If the area of the square is 144 and the perimeter of the triangle is 28, find the perimeter of the park.

   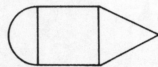

   (A) $52 + 12\pi$
   (B) $52 + 6\pi$
   (C) $40 + 6\pi$
   (D) $34 + 12\pi$
   (E) $32 + 6\pi$

9. An oil tank has a capacity of 45 gallons. At the beginning of October it is 80% full. At the end of October it is $\frac{1}{3}$ full. How many gallons of oil were used in October?

   (A) 21
   (B) 25
   (C) 41
   (D) 27
   (E) 30

10. $\overline{AB}$ and $\overline{CD}$ are diameters of circle O. The number of degrees in angle CAB is

    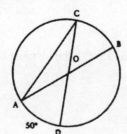

    (A) 50
    (B) 100
    (C) 130
    (D) $12\frac{1}{2}$
    (E) 25

11. If $\frac{a}{b} \cdot \frac{b}{c} \cdot \frac{c}{d} \cdot \frac{d}{e} \cdot x = 1$, then x must equal

    (A) $\frac{a}{e}$
    (B) $\frac{e}{a}$
    (C) e
    (D) $\frac{1}{a}$
    (E) none of these

12. If the sum of x and y is z and the average of m, n, and p is q, find the value of $x + y + m + n + p$ in terms of z and q.

    (A) $2z + 3q$
    (B) $z + 3q$
    (C) $z + z + \frac{q}{3}$
    (D) $\frac{z}{2} + \frac{q}{3}$
    (E) none of these

13. Isosceles triangle ABC is inscribed in square $BCDE$ as shown. If the area of square $BCDE$ is 4, the perimeter of triangle ABC is

    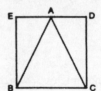

    (A) 8
    (B) $2 + \sqrt{5}$
    (C) $2 + 2\sqrt{5}$
    (D) $2 + \sqrt{10}$
    (E) 12

14. If a is not 0 or 1, a fraction equivalent to $\dfrac{\frac{1}{a}}{2 - \frac{2}{a}}$ is

    (A) $\frac{1}{2a - 2}$
    (B) $\frac{2}{a - 2}$
    (C) $\frac{1}{a - 2}$
    (D) $\frac{1}{a}$
    (E) $\frac{2}{2a - 1}$

15. At 3:30 P.M. the angle between the hands of a clock is

    (A) $90°$
    (B) $80°$
    (C) $75°$
    (D) $72°$
    (E) $65°$

16. A clerk's weekly salary is $320 after a 25% raise. What was his weekly salary before the raise?

    (A) $256
    (B) $260
    (C) $300
    (D) $304
    (E) $316

17. The figure below is composed of 5 equal squares. If the area of the figure is 125, find its perimeter.

    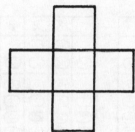

    (A) 60
    (B) 100
    (C) 80
    (D) 75
    (E) 20

18. Which of the following is equal to $\frac{1}{2}$ of $\frac{3}{5}$?

    (A) 3%
    (B) $33\frac{1}{3}$%
    (C) 30%
    (D) $83\frac{1}{3}$%
    (E) 120%

19. The length of an arc of a circle is equal to $\frac{1}{5}$ of the circumference of the circle. If the length of the arc is 2π, the radius of the circle is

    (A) 2
    (B) 1
    (C) 10
    (D) 5
    (E) $\sqrt{10}$

20. If two sides of a triangle are 3 and 4 and the third side is x, then

    (A) $x = 5$
    (B) $x > 7$
    (C) $x < 7$
    (D) $1 < x < 7$
    (E) $x > 7$ or $x < 1$

21. The smallest integer that, when squared, is less than 5 is

    (A) 0
    (B) 1
    (C) 2
    (D) 3
    (E) none of these

22. Mr. Prince takes his wife and two children to the circus. If the price of a child's ticket is $\frac{1}{2}$ the price of an adult ticket and Mr. Prince pays a total of $12.60, find the price of a child's ticket.

    (A) $4.20
    (B) $3.20
    (C) $1.60
    (D) $2.10
    (E) $3.30

23. If $\begin{pmatrix} a \\ b \ c \end{pmatrix}$ is defined as being equal to $ab - c$, then $\begin{pmatrix} 3 \\ 4 \ 5 \end{pmatrix} + \begin{pmatrix} 5 \\ 6 \ 7 \end{pmatrix}$ is equal to

    (A) 30
    (B) 40
    (C) 11
    (D) 6
    (E) 15

24. The diameter of a circle is increased by 50%. The area is increased by

    (A) 50%
    (B) 100%
    (C) 125%
    (D) 200%
    (E) 250%

25. Of the students at South High, $\frac{1}{3}$ are seniors. Of the seniors, $\frac{3}{4}$ will go to college next year. What percent of the students at South High will go to college next year?

    (A) 75
    (B) 25
    (C) $33\frac{1}{3}$
    (D) 50
    (E) 45

Section 2

25 Questions

Time: 30 Minutes

Directions: Solve each of the following problems. Write the answer in the corresponding grid on the answer sheet and fill in the ovals beneath each answer you write. Here are some examples.

Answer: 3/4 (−.75; show answer either way)   *Answer : 325*

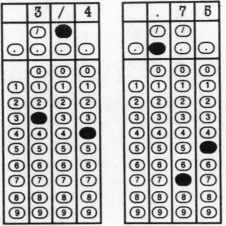   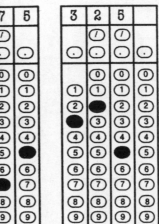   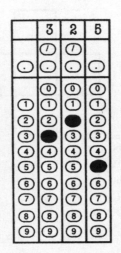

Note: A mixed number such as 3 1/2 must be gridded as 7/2 or as 3.5. If gridded as "3 1/2," it will be read as "thirty–one halves."

Note: Either position is correct.

1. If $a = 4$, what is the value of $\sqrt{a^2 + 9}$?

2. When a certain number is divided by 2, there is no remainder. If there is a remainder when the number is divided by 4, what must the remainder be?

3. If $a = x^2$ and $x = \sqrt{8}$, what is the value of a?

4. If $\frac{2}{5}x = \frac{5}{2}y$, what is the value of $\frac{y}{x}$?

5. If there are 30 students at a meeting of the Forum Club, and 20 are wearing white, 17 are wearing black and 14 are wearing both black and white, how many are wearing neither black nor white?

6. If $a \square b$ means $a \cdot b + (a - b)$, find the value of $4 \square 2$.

7. A drawer contains 4 red socks and 4 blue socks. Find the least number of socks that must be drawn from the drawer to be assured of having a pair of red socks.

8. How many 2-inch squares are needed to fill a border around the edge of the shaded square with a side of 6" as shown in the figure below?

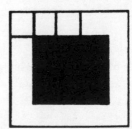

9. If $3x + 3x - 3x = 12$, what is the value of $3x + 1$?

10. If $ab = 10$ and $a^2 + b^2 = 30$, what is the value of $(a + b)^2$?

PRACTICE TEST B

Answer Sheet

Directions: For each question, darken the oval that corresponds to your answer choice. Mark only one oval for each question. If you change your mind, erase your answer completely.

Section 1

1. Ⓐ Ⓑ Ⓒ Ⓓ Ⓔ   8. Ⓐ Ⓑ Ⓒ Ⓓ Ⓔ   15. Ⓐ Ⓑ Ⓒ Ⓓ Ⓔ   22. Ⓐ Ⓑ Ⓒ Ⓓ Ⓔ
2. Ⓐ Ⓑ Ⓒ Ⓓ Ⓔ   9. Ⓐ Ⓑ Ⓒ Ⓓ Ⓔ   16. Ⓐ Ⓑ Ⓒ Ⓓ Ⓔ   23. Ⓐ Ⓑ Ⓒ Ⓓ Ⓔ
3. Ⓐ Ⓑ Ⓒ Ⓓ Ⓔ   10. Ⓐ Ⓑ Ⓒ Ⓓ Ⓔ   17. Ⓐ Ⓑ Ⓒ Ⓓ Ⓔ   24. Ⓐ Ⓑ Ⓒ Ⓓ Ⓔ
4. Ⓐ Ⓑ Ⓒ Ⓓ Ⓔ   11. Ⓐ Ⓑ Ⓒ Ⓓ Ⓔ   18. Ⓐ Ⓑ Ⓒ Ⓓ Ⓔ   25. Ⓐ Ⓑ Ⓒ Ⓓ Ⓔ
5. Ⓐ Ⓑ Ⓒ Ⓓ Ⓔ   12. Ⓐ Ⓑ Ⓒ Ⓓ Ⓔ   19. Ⓐ Ⓑ Ⓒ Ⓓ Ⓔ
6. Ⓐ Ⓑ Ⓒ Ⓓ Ⓔ   13. Ⓐ Ⓑ Ⓒ Ⓓ Ⓔ   20. Ⓐ Ⓑ Ⓒ Ⓓ Ⓔ
7. Ⓐ Ⓑ Ⓒ Ⓓ Ⓔ   14. Ⓐ Ⓑ Ⓒ Ⓓ Ⓔ   21. Ⓐ Ⓑ Ⓒ Ⓓ Ⓔ

Section 2

Note: Only the answers entered on the grid are scored. Handwritten answers at the top of the column are *not* scored.

PRACTICE TEST B

Section 1

25 Questions

Time: 30 Minutes

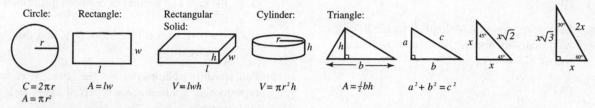

Circle: $C = 2\pi r$ $A = \pi r^2$

Rectangle: $A = lw$

Rectangular Solid: $V = lwh$

Cylinder: $V = \pi r^2 h$

Triangle: $A = \frac{1}{2}bh$ $a^2 + b^2 = c^2$

The number of degrees of arc in a circle is 360.
The measure in degrees of a straight angle is 180.
The sum of the measures in degrees of the angles of a triangle is 180.

1. A musical instrument depreciates by 20% of its value each year. What is the value, after 2 years, of a piano purchased new for $1200?

    (A) $768
    (B) $912
    (C) $675
    (D) $48
    (E) $1152

2. Which of the following has the largest numerical value?

    (A) $\dfrac{3}{5}$

    (B) $\left(\dfrac{2}{3}\right)\left(\dfrac{3}{4}\right)$

    (C) $\sqrt{.25}$

    (D) $(.9)^2$

    (E) $\dfrac{2}{.3}$

3. $\dfrac{1}{4}\%$ written as a decimal is

    (A) 25
    (B) 2.5
    (C) .25
    (D) .025
    (E) .0025

4. Which of the following fractions is equal to $\dfrac{1}{4}\%$?

    (A) $\dfrac{1}{25}$

    (B) $\dfrac{4}{25}$

    (C) $\dfrac{1}{4}$

    (D) $\dfrac{1}{400}$

    (E) $\dfrac{1}{40}$

5. Roger receives a basic weekly salary of $80 plus a 5% commission on his sales. In a week in which his sales amounted to $800, the ratio of his basic salary to his commission was

    (A) 2:1
    (B) 1:2
    (C) 2:3
    (D) 3:2
    (E) 3:1

6. The value of $\dfrac{\frac{1}{2}}{\frac{1}{3} - \frac{1}{4}}$ is

    (A) 6

    (B) $\dfrac{1}{6}$

    (C) 1

    (D) 3

    (E) $\dfrac{3}{2}$

7. The sum of Alan's age and Bob's age is 40. The sum of Bob's age and Carl's age is 34. The sum of Alan's age and Carl's age is 42. How old is Bob?

   (A) 18
   (B) 24
   (C) 20
   (D) 16
   (E) 12

8. On a map having a scale of $\frac{1}{4}$ inch : 20 miles, how many inches should there be between towns 325 miles apart?

   (A) $4\frac{1}{16}$
   (B) $16\frac{1}{4}$
   (C) $81\frac{1}{4}$
   (D) $32\frac{1}{2}$
   (E) $6\frac{1}{4}$

9. In Simon's General Score, there are m male employees and f female employees. What part of the staff is men?

   (A) $\frac{m+f}{m}$
   (B) $\frac{m+f}{f}$
   (C) $\frac{m}{f}$
   (D) $\frac{m}{m+f}$
   (E) $\frac{f}{m}$

10. If the angles of a triangle are in the ratio 2:3:4, the triangle is

    (A) acute
    (B) isosceles
    (C) right
    (D) equilateral
    (E) obtuse

11. If the length and width of a rectangle are each multiplied by 2, then

    (A) the area and perimeter are both multiplied by 4
    (B) the area is multiplied by 2 and the perimeter by 4
    (C) the area is multiplied by 4 and the perimeter by 2
    (D) the area and perimeter are both multiplied by 2
    (E) the perimeter is multiplied by 4 and the area by 8

12. Paul needs m minutes to mow the lawn. After he works for k minutes, what part of the lawn is still unmowed?

    (A) $\frac{k}{m}$
    (B) $\frac{m}{k}$
    (C) $\frac{m-k}{k}$
    (D) $\frac{m-k}{m}$
    (E) $\frac{k-m}{m}$

13. Mr. Marcus earns $250 per week. If he spends 20% of his income for rent, 25% for food, and 10% for savings, how much is left each week for other expenses?

    (A) $112.50
    (B) $125
    (C) $137.50
    (D) $132.50
    (E) $140

14. What is the area of the shaded portion if the perimeter of the square is 32? (The four circles are tangent to each other and the square, and are congruent.)

    (A) $32 - 16\pi$
    (B) $64 - 16\pi$
    (C) $64 - 64\pi$
    (D) $64 - 8\pi$
    (E) $32 - 4\pi$

15. How far is the point (−3, −4) from the origin?

    (A) 2
    (B) 2.5
    (C) $4\sqrt{2}$
    (D) $4\sqrt{3}$
    (E) 5

16. The product of 3456 and 789 is exactly

    (A) 2726787
    (B) 2726785
    (C) 2726781
    (D) 2726784
    (E) 2726786

17. Susan got up one morning at 7:42 A.M. and went to bed that evening at 10:10 P.M. How much time elapsed between her getting up and going to bed that day?

    (A) 18 hrs. 2 min.
    (B) 14 hrs. 18 min.
    (C) 15 hrs. 18 min.
    (D) 9 hrs. 22 min.
    (E) 14 hrs. 28 min.

18. Find the perimeter of right triangle *ABC* if the area of square *AEDC* is 100 and the area of square *BCFG* is 36.

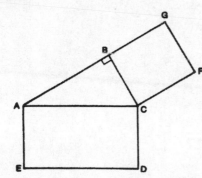

    (A) 22
    (B) 24
    (C) $16 + 6\sqrt{3}$
    (D) $16 + 6\sqrt{2}$
    (E) cannot be determined from information given

19. Find the number of degrees in angle 1 if *AB* = *AC*, *DE* = *DC*, angle 2 = 40°, and angle 3 = 80°.

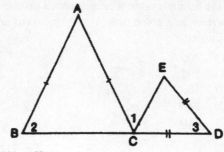

    (A) 60
    (B) 40
    (C) 90
    (D) 50
    (E) 80

20. If *p* pencils cost 2*D* dollars, how many pencils can be bought for *c* cents?

    (A) $\dfrac{pc}{2D}$
    (B) $\dfrac{pc}{200D}$
    (C) $\dfrac{50pc}{D}$
    (D) $\dfrac{2Dp}{c}$
    (E) $200pcD$

21. Two trains start from the same station at 10 A.M., one traveling east at 60 m.p.h. and the other west at 70 m.p.h. At what time will they be 455 miles apart?

    (A) 3:30 P.M.
    (B) 12:30 P.M.
    (C) 1:30 P.M.
    (D) 1 P.M.
    (E) 2 P.M.

22. If *x* < 0 and *y* < 0, then

    (A) $x + y > 0$
    (B) $x = -y$
    (C) $x > y$
    (D) $xy > 0$
    (E) $xy < 0$

23. Which of the following is the product of 4327 and 546?

    (A) 2362541
    (B) 2362542
    (C) 2362543
    (D) 2362546
    (E) 2362548

24. If a classroom contains 20 to 24 students and each corridor contains 8 to 10 classrooms, what is the minimum number of students on one corridor at a given time, if all classrooms are occupied?

    (A) 200
    (B) 192
    (C) 160
    (D) 240
    (E) 210

25. If the area of each circle enclosed in rectangle *ABCD* is 9π, the area of *ABCD* is

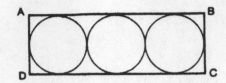

    (A) 108
    (B) 27
    (C) 54
    (D) 54π
    (E) 108π

Section 2

25 Questions

Time: 30 Minutes

> **Directions: Solve each of the following problems. Write the answer in the corresponding grid on the answer sheet and fill in the ovals beneath each answer you write. Here are some examples**

Answer: 3/4 (–.75; show answer either way)   *Answer : 325*

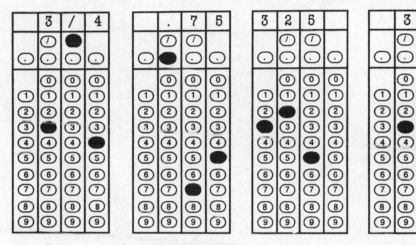

Note: A mixed number such as 3 1/2 must be gridded as 7/2 or as 3.5. If gridded as "3 1/2," it will be read as "thirty–one halves."

Note: Either position is correct.

1. Simplified as a fraction to simplest form, what part of a dime is a quarter?

2. Marion is paid $24 for 5 hours of work in the school office. Janet works 3 hours and makes $10.95. How much more per hour does Marion make than Janet?

3. If the outer diameter of a cylindrical oil tank is 54.28 inches and the inner diameter is 48.7 inches, what is the thickness of the wall of the tank, in inches?

4. What number added to 40% of itself is equal to 84?

5. If $r = 25 - s$, what is the value of $4r + 4s$?

6. A plane flies over Denver at 11:20 A.M. It passes over Coolidge, 120 miles from Denver, at 11:32 A.M. Find the rate of the plane in miles per hour.

7. 53% of the 1000 students at Jackson High are girls. How many boys are there in the school?

8. How many digits are there in the square root of a perfect square of 12 digits?

9. In May, Carter's Appliances sold 40 washing machines. In June, because of a special promotion, the store sold 80 washing machines. What is the percent of increase in the number of washing machines sold?

10. Find the value of $\left(3\sqrt{2}\right)^2$.

PRACTICE TEST C
Answer Sheet

Directions: For each question, darken the oval that corresponds to your answer choice. Mark only one oval for each question. If you change your mind, erase your answer completely.

Section 1

1. Ⓐ Ⓑ Ⓒ Ⓓ Ⓔ
2. Ⓐ Ⓑ Ⓒ Ⓓ Ⓔ
3. Ⓐ Ⓑ Ⓒ Ⓓ Ⓔ
4. Ⓐ Ⓑ Ⓒ Ⓓ Ⓔ
5. Ⓐ Ⓑ Ⓒ Ⓓ Ⓔ
6. Ⓐ Ⓑ Ⓒ Ⓓ Ⓔ
7. Ⓐ Ⓑ Ⓒ Ⓓ Ⓔ

8. Ⓐ Ⓑ Ⓒ Ⓓ Ⓔ
9. Ⓐ Ⓑ Ⓒ Ⓓ Ⓔ
10. Ⓐ Ⓑ Ⓒ Ⓓ Ⓔ
11. Ⓐ Ⓑ Ⓒ Ⓓ Ⓔ
12. Ⓐ Ⓑ Ⓒ Ⓓ Ⓔ
13. Ⓐ Ⓑ Ⓒ Ⓓ Ⓔ
14. Ⓐ Ⓑ Ⓒ Ⓓ Ⓔ

15. Ⓐ Ⓑ Ⓒ Ⓓ Ⓔ
16. Ⓐ Ⓑ Ⓒ Ⓓ Ⓔ
17. Ⓐ Ⓑ Ⓒ Ⓓ Ⓔ
18. Ⓐ Ⓑ Ⓒ Ⓓ Ⓔ
19. Ⓐ Ⓑ Ⓒ Ⓓ Ⓔ
20. Ⓐ Ⓑ Ⓒ Ⓓ Ⓔ
21. Ⓐ Ⓑ Ⓒ Ⓓ Ⓔ

22. Ⓐ Ⓑ Ⓒ Ⓓ Ⓔ
23. Ⓐ Ⓑ Ⓒ Ⓓ Ⓔ
24. Ⓐ Ⓑ Ⓒ Ⓓ Ⓔ
25. Ⓐ Ⓑ Ⓒ Ⓓ Ⓔ

Section 2

Note: Only the answers entered on the grid are scored. Handwritten answers at the top of the column are *not* scored.

PRACTICE TEST C

Section 1

25 Questions

Time: 30 Minutes

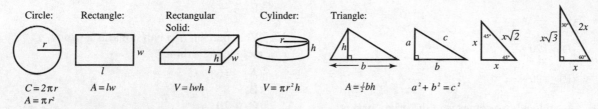

The number of degrees of arc in a circle is 360.
The measure in degrees of a straight angle is 180.
The sum of the measures in degrees of the angles of a triangle is 180.

1. $8 \cdot 8 = 4^x$. Find x

   (A) 2
   (B) 3
   (C) 4
   (D) 5
   (E) 6

2. If $a > 2$, which of the following is the smallest?

   (A) $\dfrac{2}{a}$

   (B) $\dfrac{a}{2}$

   (C) $\dfrac{a+1}{2}$

   (D) $\dfrac{2}{a+1}$

   (E) $\dfrac{2}{a-1}$

3. Which of the following has the greatest value?

   (A) $\dfrac{1}{2}$

   (B) $\sqrt{.2}$

   (C) $.2$

   (D) $(.2)^2$

   (E) $(.02)^3$

4. If $\dfrac{a}{b} = \dfrac{3}{4}$, then $12a =$

   (A) $3b$
   (B) b
   (C) $9b$
   (D) $12b$
   (E) $16b$

5. If $a = b$ and $\dfrac{1}{c} = b$, then $c =$

   (A) a
   (B) $-a$
   (C) b
   (D) $\dfrac{1}{a}$
   (E) $-b$

6. If a building B feet high casts a shadow F feet long, then, at the same time of day, a tree T feet high will cast a shadow how many feet long?

   (A) $\dfrac{FT}{B}$

   (B) $\dfrac{FB}{T}$

   (C) $\dfrac{B}{FT}$

   (D) $\dfrac{TB}{F}$

   (E) $\dfrac{T}{FB}$

7. The vertices of a triangle are (3,1) (8,1) and (8,3). The area of this triangle is

   (A) 5
   (B) 10
   (C) 7
   (D) 20
   (E) 14

8. Of 60 employees at the Star Manufacturing Company, *x* employees are female. If $\frac{2}{3}$ of the remainder are married, how many unmarried men work for this company?

   (A) $40 - \frac{2}{3}x$

   (B) $40 - \frac{1}{3}x$

   (C) $40 + \frac{1}{3}x$

   (D) $20 - \frac{2}{3}x$

   (E) $20 - \frac{1}{3}x$

9. A circle whose center is at the origin passes through the point whose coordinates are (1,1). The area of the circle is

   (A) π
   (B) 2π
   (C) $\sqrt{2\pi}$
   (D) $2\sqrt{2\pi}$
   (E) 4π

10. In triangle *ABC*, *AB = BC* and *AC* is extended to *D*. If angle *BCD* contains 100°, find the number of degrees in angle *B*.

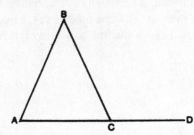

   (A)   50
   (B)   80
   (C)   60
   (D)   40
   (E)   20

11. $\dfrac{4\frac{1}{2}}{10\frac{1}{8}}$

   (A) $\frac{2}{5}$

   (B) $\frac{4}{9}$

   (C) $\frac{4}{81}$

   (D) $\frac{3}{7}$

   (E) $\frac{15}{23}$

12. Which of the following is greater than $\frac{1}{3}$?

   (A)   .33

   (B) $\left(\dfrac{1}{3}\right)^2$

   (C) $\dfrac{1}{4}$

   (D) $\dfrac{1}{.3}$

   (E) $\dfrac{.3}{2}$

13. What percent of a half dollar is a penny, a nickel, and a dime?

   (A)   16
   (B)   8
   (C)   20
   (D)   25
   (E)   32

14. If $\dfrac{1}{a} + \dfrac{1}{b} = \dfrac{1}{c}$ then $c =$

   (A) $a + b$
   (B) ab

   (C) $\dfrac{a+b}{ab}$

   (D) $\dfrac{ab}{a+b}$

   (E) $\dfrac{1}{2}ab$

15. What percent of *a* is *b*?

   (A) $\dfrac{100b}{a}$

   (B) $\dfrac{a}{b}$

   (C) $\dfrac{b}{100a}$

   (D) $\dfrac{b}{a}$

   (E) $\dfrac{100a}{b}$

16. The average of two numbers is *A*. If one of the numbers is *x*, the other number is

   (A) $A - x$

   (B) $\dfrac{A}{2} - x$

   (C) $2A - x$

   (D) $\dfrac{A+x}{2}$

   (E) $x - A$

17. If $a = 5b$, then $\dfrac{3}{5}a =$

    (A) $\dfrac{5b}{3}$

    (B) $3b$

    (C) $\dfrac{3b}{5}$

    (D) $\dfrac{b}{3}$

    (E) $\dfrac{b}{5}$

18. A rectangular door measures 5 feet by 6 feet 8 inches. The distance from one corner of the door to the diagonally opposite corner is

    (A) $9'4''$
    (B) $8'4''$
    (C) $8'3''$
    (D) $9'6''$
    (E) $9'$

19. Two ships leave from the same port at 11:30 A.M. If one sails due east at 20 miles per hour and the other due south at 15 miles per hour, how many miles apart are the ships at 2:30 P.M.?

    (A) 25
    (B) 50
    (C) 75
    (D) 80
    (E) 35

20. If m men can paint a house in d days, how many days will it take $m + 2$ men to paint the same house?

    (A) $d + 2$
    (B) $d - 2$
    (C) $\dfrac{m+2}{md}$
    (D) $\dfrac{md}{m+2}$
    (E) $\dfrac{md + 2d}{m}$

21. Ken received grades of 90, 88, and 75 on three tests. What grade must he receive on the next test so that his average for these 4 tests is 85?

    (A) 87
    (B) 92
    (C) 83
    (D) 85
    (E) 88

22. There is enough food at a picnic to feed 20 adults or 32 children. If there are 15 adults at the picnic, how many children can still be fed?

    (A) 10
    (B) 8
    (C) 16
    (D) 12
    (E) 4

23. In parallelogram $ABCD$, angle A contains 60°. The sum of angle B and angle D must be

    (A) 120°
    (B) 300°
    (C) 240°
    (D) 60°
    (E) 180°

24. The area of circle O is 64π. The perimeter of square $ABCD$ is

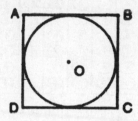

    (A) 32
    (B) 32π
    (C) 64
    (D) 16
    (E) 64π

25. If a train covers 14 miles in 10 minutes, then the rate of the train in miles per hour is

    (A) 140
    (B) 112
    (C) 84
    (D) 100
    (E) 98

Section 2

25 Questions

Time: 30 Minutes

Directions: Solve each of the following problems. Write the answer in the corresponding grid on the answer sheet and fill in the ovals beneath each answer you write. Here are some examples.

Answer: 3/4 (–.75; show answer either way)   *Answer : 325*

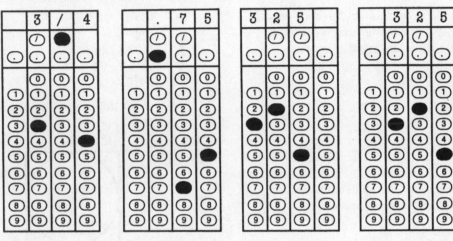

Note: A mixed number such as 3 1/2 must be gridded as 7/2 or as 3.5. If gridded as "3 1/2," it will be read as "thirty–one halves."

Note: Either position is correct.

1. If $\frac{8}{8}$ of $\frac{3}{8}$ is added to $\frac{3}{8}$, what is the result?

2. If $2^{n-3} = 32$ what is the value of n?

3. In a group of 40 students, 25 applied to Columbia and 30 applied to Cornell. If 3 students applied to neither Columbia nor Cornell, how many students applied to both schools?

4. If $x^2 - y^2 = 100$ and $x - y = 20$, what is the value of $x + y$?

5. A gallon of water is added to 6 quarts of a solution that is 50% acid. What percent of the new solution is acid?

6. A gasoline tank is $\frac{1}{4}$ full. After adding 10 gallons of gasoline, the gauge indicates that the tank is $\frac{2}{3}$ full. Find the capacity of the tank in gallons.

7. If $(x - y)^2 = 40$ and $x^2 + y^2 = 60$, what is the value of xy?

8. If 2.5 cm = 1 in. and 36 in. = 1 yd., how many centimeters are in 1 yard?

9. How much more is $\frac{1}{4}$ of $\frac{1}{3}$ than $\frac{1}{3}$ of $\frac{1}{4}$?

10. If the average of 5 consecutive even integers is 82, what is the largest of these integers?

SOLUTIONS TO PRACTICE TESTS

PRACTICE TEST A

Section 1

1. **(B)** $\dfrac{1}{5}x = 8$

   $x = 40$

   $\dfrac{1}{4}(40) = 10$

2. **(E)** Multiples of 3 are 3 apart x is 3 below $x + 3$. $x + 6$ is 3 above $x + 3$. $6x + 18 = 6(x + 3)$, $2x + 6 = 2(x + 3)$. $3x + 5$ does not have a factor of 3, nor can it be shown to differ front $x + 3$ by a multiple of 3.

3. **(C)** Angle $C = 40°$ (Congruent angles.)

   Angle $BAC = 100°$ (Sum of the angles in a triangle is 180°.)

   Angle $x = 100°$ (Vertical angles are congruent.)

4. **(E)**

   $\dfrac{2}{5} \cdot \dfrac{3}{2} = \dfrac{3}{5}$

   $\dfrac{1}{2} - \dfrac{1}{10} = \dfrac{10 - 2}{20} = \dfrac{8}{20} = \dfrac{2}{5}$

   $\dfrac{3}{5} + \dfrac{2}{5} = 1$

5. **(C)** Basic toll $1.00.

   Extra toll $2.25, which is 3($.75).

   Therefore, the car holds a driver and 3 extra passengers, for a total of 4 persons.

6. **(C)** Divide by y^2 : $y = 2$.

7. **(C)** $x = y + 1$

   Using the largest negative integers will give the smallest product. Let $y = -2$, $x = -1$, then $xy = 2$.

8. **(C)** Side of square $= 12 =$ diameter of semicircle.

   Remaining 2 sides of triangle add up to 16.

   Perimeter of semicircle $= \dfrac{1}{2}\pi d = \dfrac{1}{2} \cdot \pi \cdot 12 = 6\pi$

   2 sides of square in perimeter $= 24$

   Total perimeter of park $= 16 + 6\pi + 24 = 40 + 6\pi$

9. **(A)** $80\% = \dfrac{4}{5}$ $\dfrac{4}{5} \cdot 45 = 36$

   $\dfrac{1}{3} \cdot 45 = 15$

   Used in October $= 36 - 15 = 21$

10. **(E)** Angle $AOD = 50°$

    Angle $COB = 50°$

    Arc $CB = 50°$

    Angle CAB is an inscribed angle $= 25°$

11. **(B)**

    $\dfrac{a}{\not b} \cdot \dfrac{\not b}{\not c} \cdot \dfrac{\not c}{\not d} \cdot \dfrac{\not d}{e} \cdot x = 1$

    $\dfrac{a}{e} \cdot x = 1$

    $x = \dfrac{e}{a}$

12. **(B)**

    $\dfrac{m + n + p}{3} = q$

    $m + n + p = 3q$

    $\dfrac{x + y = z}{m + n + p + x + y = 3q + z}$

13. **(C)** Side of square $= 2$

    If $BE = 2$, $EA = 1$, then by the Pythagorean theorem, BA and AC each equal $\sqrt{5}$.

    Perimeter of triangle $ABC = 2 + 2\sqrt{5}$.

14. **(A)** Multiply every term by a.

    $\dfrac{1}{2a - 2}$

15. **(C)** There are 30° in each of the 12 even spaces between numbers on the clock. At 3:30, the minute hand points to 6 and the hour hand is halfway between 3 and 4. The angle between the hands is $2\dfrac{1}{2}(30°) = 75°$.

16. **(A)** $320 is 125% of his former salary.

    $320 = 1.25x$

    $32000 = 125x$

    $256 = x$

17. (A) Area of each square $= \frac{1}{5} \cdot 125 = 25$

Side of each square = 5

Perimeter is made up of 12 sides. 12(5) = 60

18. (C) $\frac{1}{2} \cdot \frac{3}{5} = \frac{3}{10} = 30\%$

19. (D) Circumference is 5 times arc.

$5(2\pi) = 10\pi = \pi d$

$d = 10 \qquad r = 5$

20. (D) The sum of any two sides of a triangle must be greater than the third side. Therefore, x must be less than 7 ($4 + 3 > x$); however, x must be greater than 1, as $3 + x > 4$.

21. (E) x can be negative as $(-2)^2 = 4$, which is less than 5.

22. (D) The two children's tickets equal one adult ticket. Mr. Prince pays the equivalent of 3 adult tickets.

$3a = 12.60$

$a = 4.20$

Child's ticket $= \frac{1}{2}(4.20) = \$2.10$

23. (A)

$\begin{pmatrix} & 3 & \\ 4 & & 5 \end{pmatrix} = 12 - 5 = 7$

$\begin{pmatrix} & 5 & \\ 6 & & 7 \end{pmatrix} = 30 - 7 = 23$

$7 + 23 = 30$

24. (C) If the linear ratio is 1:1.5, then the area ratio is $(1)^2 : (1.5)^2$ or 1:2.25. The increase is 1.25 or 125% of the original area.

25. (B) $\frac{3}{4}$ of $\frac{1}{3}$ will go on to college next year.

$\frac{3}{4} \cdot \frac{1}{3} = \frac{1}{4} = 25\%$.

Section 2

1. $\sqrt{4^2 + 9} = \sqrt{25} = 5$ **(answer)**

2. The number must be an even number, as there is no remainder when divided by 2. If division by 4 does give a remainder, it must be 2, since even numbers are 2 apart. 2 **(answer)**

3. $\left(\sqrt{8}\right)^2 = 8$ **(answer)**

4. $\dfrac{y}{x} = \dfrac{\frac{2}{5}}{\frac{5}{2}} = \dfrac{2}{5} \cdot \dfrac{2}{5} = \dfrac{4}{25}$ **(answer)**

5. Illustrate the given facts as follows.

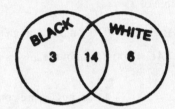

This accounts for 23 students, leaving 7. **(answer)**

6. $4 \,\square\, 2 = 4 \cdot 2 + (4 - 2) = 8 + 2 = 10$ **(answer)**

7. It is possible for the first four to be blue, but then the next two must be red. Of course it is possible that two red socks could be drawn earlier, but with 6 we are *assured* of a pair of red socks. 6 **(answer)**

8. 16 **(answer)**

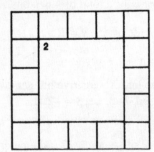

9. $3x = 12$

$x = 4$

$3x + 1 = 13$ **(answer)**

10. $\left(a+b\right)^2 = a^2 + 2ab + b^2$

$a^2 + b^2 = 30$

$\underline{\quad\quad 2ab = 20\quad}$

$a^2 + 2ab + b^2 = 50$ **(answer)**

PRACTICE TEST B

Section 1

1. **(A)** $20\% = \dfrac{1}{5}$

   $\dfrac{1}{5} \cdot 1200 = \$240$ depreciation first year.

   $\$1200 - \$240 = \$960$ value after 1 year.

   $\dfrac{1}{5} \cdot 960 = \$192$ depreciation second year.

   $\$960 - \$192 = \$768$ value after 2 years.

2. **(E)**

   $\dfrac{3}{5} = .6$

   $\left(\dfrac{2}{3}\right)\left(\dfrac{3}{4}\right) = \dfrac{1}{2} = .5$

   $\sqrt{.25} = .5$

   $(.9)^2 = .81$

   $\dfrac{2}{.3} = \dfrac{20}{3} = 6.\overline{6}$

3. **(E)**

   $\dfrac{1}{4} = .25$

   $\dfrac{1}{4}\% = .25\% = .0025$

4. **(D)** $\dfrac{1}{4}\% = \dfrac{1}{4} \div 100 = \dfrac{1}{4} \cdot \dfrac{1}{100} = \dfrac{1}{400}$

5. **(A)** $.05\,(800) = \$40$ commission

   $80{:}40 = 2{:}1$

6. **(A)** Multiply every term by 12.

   $\dfrac{6}{4-3} = 6$

7. **(D)** $A + B = 40$
   $\quad\quad\ B + C = 34$
   $\quad\quad\ A + C = 42$

   Subtract second equation from third.

   $A - B = 8$

   Subtract from first equation.

   $2B = 32$

   $B = 16$

8. **(A)** Use a proportion comparing inches to miles.

   $\dfrac{\frac{1}{4}}{20} = \dfrac{x}{325}$

   $20x = \dfrac{325}{4}$

   $x = \dfrac{325}{4} \cdot \dfrac{1}{20} = \dfrac{325}{80} = 4\dfrac{5}{80} = 4\dfrac{1}{16}$

9. **(D)** There are $m + f$ people on the staff. Of these, m are men.

   $\dfrac{m}{m+f}$ of the staff is men.

10. **(A)** Represent the angles as $2x$, $3x$, and $4x$.
    $9x = 180$
    $x = 20$

    The angles are $40°$, $60°$, and $80°$, all acute.

11. **(C)** The linear ratio stays constant, so the perimeter is also multiplied by 2. The area ratio is the square of the linear ratio, so the area is multiplied by 2^2 or 4.

12. **(D)** In k minutes, $\dfrac{k}{m}$ of the lawn is mowed.

    Still undone is $1 - \dfrac{k}{m}$ or $\dfrac{m-k}{m}$

13. **(A)** 55% of his salary is spent. 45% is left.

    There is only one answer among the choices less than $\dfrac{1}{2}$ of his salary.

14. **(B)** Each side of square $= 8$
    Radius circle $= 2$
    Area of square $= 8^2 = 64$
    Area of 4 circles $= 4\pi\,r^2 = 4 \cdot \pi \cdot 2^2 = 16\pi$
    Shaded area $= 64 - 16\pi$

15. **(E)** Plotting the point shows a 3, 4, 5 triangle.

    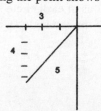

16. **(D)** Since 6 times 9 is 54, the product must end in 4.

17. (E)   Figure the time elapsed on either side of 12 noon. From 7:42 A.M. to 12 noon is 4 hrs. 18 min. From 12 noon, to 10:10 P.M. is 10 hrs. 10 min. The sum of the two is 14 hrs. 28 min.

18. (B)   Each side of square *AEDC* is 10.

    Each side of square *BCFG* is 6.

    Triangle *ABC* is a 6, 8, 10 triangle, making the perimeter 24.

19. (C)   There are 90° left for angle 1 since angle *BCD* is a straight angle.

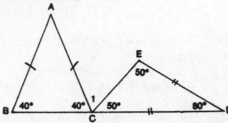

20. (B)   Use a proportion comparing pencils to cents. Change 2*D* dollars to 200*D* cents.

$$\frac{p}{200D} = \frac{x}{c}$$

$$\frac{pc}{200D} = x$$

21. (C)   Distance of first train = 60*x*

    Distance of second train = 70*x*

    $60x + 70x = 455$

    $130x = 455$

    $x = 3\frac{1}{2}$

    In $3\frac{1}{2}$ hours, the time will be 1:30 P.M.

22. (D)   When two negative numbers are multiplied, their product is positive.

23. (B)   Since 7 times 6 is 42, the product must end in 2.

24. (C)   The minimum is 20 students in 8 classrooms.

25. (A)   The radius of each circle is 3, making the dimensions of the rectangle 18 by 6, and the area (18)(6), or 108.

Section 2

1. $\frac{25}{10} = \frac{5}{2}$ **(answer)**

2.   Marion's hourly wage is $\frac{\$24}{5}$ or \$4.80 .

    Janet's hourly wage is $\frac{\$10.95}{3}$ or \$3.65.

    $4.80 – $3.65 = $1.15.   **(answer)**

3.   The difference of 5.58 must be divided between both ends. The thickness on each side is 2.79. **(answer)**

4. $x + .40x = 84$

    $1.40x = 84$

    $14x = 840$

    $x = 60$ **(answer)**

5. $r + s = 25$

    $4(r + s) = 4(25) = 100$ **(answer)**

6.   The plane covers 120 miles in 12 minutes or $\frac{1}{5}$ hour. In $\frac{5}{5}$ or 1 hour, it covers 5(120), or 600 miles. 600 **(answer)**

7.   47% of 1000 are boys.

    $(.47)(1000) = 470$ boys **(answer)**

8.   For every pair of digits in a number, there will be one digit in the square root. 6 **(answer)**

9.   Increase of 40

    $\text{Percent of Increase} = \frac{\text{Amount of increase}}{\text{Original}} \cdot 100\%$

    $\frac{40}{40} \cdot 100\% = 100\%$

    (answer)

10. $\left(3\sqrt{2}\right)\left(3\sqrt{2}\right) = 9 \cdot 2 = 18$ **(answer)**

PRACTICE TEST C

Section 1

1. **(B)** $64 = 4^x$
   $x = 3$ $\quad(4 \cdot 4 \cdot 4 = 64)$

2. **(D)** B and C are greater than 1. A, D, and E all have the same numerator. In this case, the one with the largest denominator will be the smallest fraction.

3. **(A)** $\dfrac{1}{2} = .5$ $\qquad \sqrt{.2} = .45$
   $(.2)^2 = .04 \qquad (.02)^3 = .000008$

4. **(C)** Cross multiply.
   $4a = 3b$
   Multiply by 3.
   $12a = 9b$

5. **(D)** $a = b = \dfrac{1}{c}$
   $a = \dfrac{1}{c}$
   $ac = 1$
   $c = \dfrac{1}{a}$

6. **(A)** The ratio of height to shadow is constant.
   $\dfrac{B}{F} = \dfrac{T}{x}$
   $Bx = FT$
   $x = \dfrac{FT}{B}$

7. **(A)** Right triangle area $= \dfrac{1}{2} \cdot 5 \cdot 2 = 5$

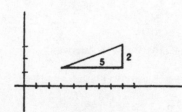

8. **(E)** $60 - x$ employees are male
   $\dfrac{1}{3}$ of these unmarried
   $\dfrac{1}{3}(60 - x) = 20 - \dfrac{1}{3}x$

9. **(B)** $1^2 + 1^2 = r^2$
   $2 = r^2$
   Area $= \pi r^2 = 2\pi$

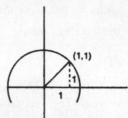

10. **(E)** Angle BCA = Angle BAC = 80°
    There are 20° left for angle B.

11. **(B)** $\dfrac{9}{2} \div \dfrac{81}{8} = \dfrac{9}{2} \cdot \dfrac{8}{81} = \dfrac{4}{9}$

12. **(D)** $\dfrac{1}{.3} = \dfrac{10}{3} = 3\dfrac{1}{3}$

13. **(E)** $\dfrac{16}{50} = \dfrac{32}{100} = 32\%$

14. **(D)** Multiply by abc.
    $bc + ac = ab$
    $c(b + a) = ab$
    $c = \dfrac{ab}{b+a}$

15. **(A)** $\dfrac{b}{a} \cdot 100 = \dfrac{100b}{a}$

16. **(C)**
    $\dfrac{x+y}{2} = A$
    $x + y = 2A$
    $y = 2A - x$

17. **(B)** $\dfrac{3}{5} \cdot 5b = 3b$

18. **(B)** 5 feet = 60 inches
    6 feet 8 inches = 80 inches
    This is a 6, 8, 10 triangle, making the diagonal 100 inches, which is 8 feet 4 inches.

19. **(C)** In 3 hours, one ship went 60 miles, the other 45 miles. This is a 3, 4, 5 triangle as 45 = 3(15), 60 = 4(15). The hypotenuse will be 5(15), or 75.

20. **(D)** This is inverse variation.

$m \cdot d = (m+2) \cdot x$

$\dfrac{md}{m+2} = x$

21. (A) He must score as many points above 85 as below. So far he has 8 above and 10 below. He needs another 2 above.

22. (B) If 15 adults are fed, $\dfrac{3}{4}$ of the food is gone. $\dfrac{1}{4}$ of the food will feed $\dfrac{1}{4} \cdot 32$, or 8, children.

23. (C) If angle $A = 60°$, then angle $B = 120°$. Angle $B =$ Angle D. Their sum is 240°.

24. (C) Area of circle $= 64\pi = r^2$

Radius of circle $= 8$

Side of square $= 16$

Perimeter of square $= 64$

25. (C) 10 minutes $= \dfrac{1}{6}$ hour

In one hour, the train will cover 6(14), or 84 miles.

Section 2

1. $\dfrac{3}{8} + \dfrac{3}{8} = \dfrac{6}{8}$ or $\dfrac{3}{4}$ **(answer; both acceptable)**

2. $2^{n-3} = 2^5$

$n - 3 = 5$

$n = 8$ **(answer)**

3. $25 - x + x + 30 - x = 37$

$55 - x = 37$

$18 = x$

18 **(answer)**

4. $x^2 - y^2 = (x-y)(x+y)$

$100 = 20(x+y)$

$5 = (x+y)$

5 **(answer)**

5. $\dfrac{3}{10} = 30\%$ **(answer)**

	No. of quarts	% acid	= Amount of acid
Original	6	.50	3
Added	4	0	0
New	10		3

6. 10 gallons is $\dfrac{2}{3} - \dfrac{1}{4}$ of the tank.

$\dfrac{2}{3} - \dfrac{1}{4} = \dfrac{8-3}{12} = \dfrac{5}{12}$

$\dfrac{5}{12}x = 10$

$5x = 120$

$x = 24$ **(answer)**

7. $(x-y)^2 = x^2 - 2xy + y^2$

$40 = 60 - 2xy$

$2xy = 20$

$xy = 10$ **(answer)**

8. $36(2.5) = 90$ **(answer)**

9. $\dfrac{1}{4} \cdot \dfrac{1}{3} = \dfrac{1}{12}$ $\dfrac{1}{3} \cdot \dfrac{1}{4} = \dfrac{1}{12}$

0 **(answer)**

10. The average is the middle integer. If 82 is the third, 86 is the last.

86 **(answer)**

Notes

Notes

Notes

Notes

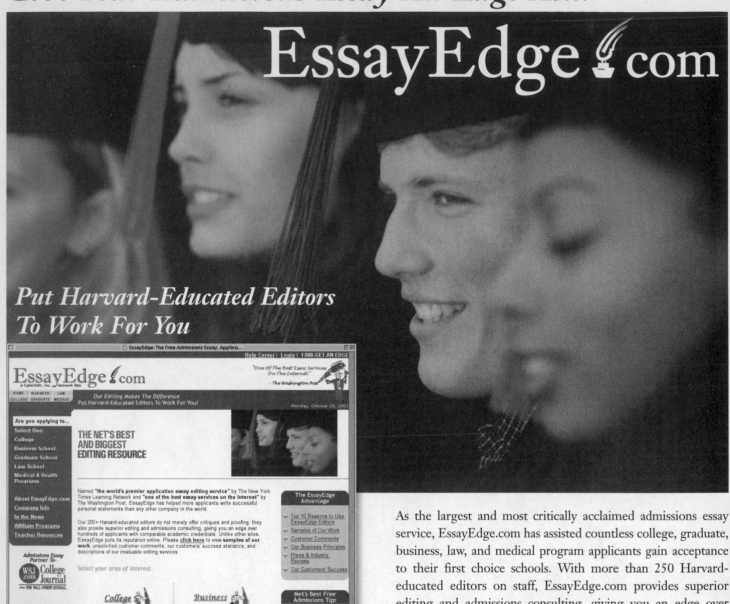